T0292588

Smart Innovation, Systems and Technologies

Volume 48

Series editors

Robert J. Howlett, KES International, Shoreham-by-sea, UK
e-mail: rjhowlett@kesinternational.org

Lakhmi C. Jain, Bournemouth University, Poole, UK, and
University of South Australia, Adelaide, Australia
e-mail: Lakhmi.jain@unisa.edu.au

About this Series

The Smart Innovation, Systems and Technologies book series encompasses the topics of knowledge, intelligence, innovation and sustainability. The aim of the series is to make available a platform for the publication of books on all aspects of single and multi-disciplinary research on these themes in order to make the latest results available in a readily-accessible form. Volumes on interdisciplinary research combining two or more of these areas is particularly sought.

The series covers systems and paradigms that employ knowledge and intelligence in a broad sense. Its scope is systems having embedded knowledge and intelligence, which may be applied to the solution of world problems in industry, the environment and the community. It also focusses on the knowledgetransfer methodologies and innovation strategies employed to make this happen effectively. The combination of intelligent systems tools and a broad range of applications introduces a need for a synergy of disciplines from science, technology, business and the humanities. The series will include conference proceedings, edited collections, monographs, handbooks, reference books, and other relevant types of book in areas of science and technology where smart systems and technologies can offer innovative solutions.

High quality content is an essential feature for all book proposals accepted for the series. It is expected that editors of all accepted volumes will ensure that contributions are subjected to an appropriate level of reviewing process and adhere to KES quality principles.

More information about this series at http://www.springer.com/series/8767

Anna Esposito · Marcos Faundez-Zanuy
Antonietta M. Esposito · Gennaro Cordasco
Thomas Drugman · Jordi Solé-Casals
Francesco Carlo Morabito
Editors

Recent Advances in Nonlinear Speech Processing

 Springer

Editors
Anna Esposito
Department of Psychology
Seconda Università di Napoli and IIASS
Caserta
Italy

Marcos Faundez-Zanuy
Escola Superior Politècnica Tecnocampus
  (Pompeu Fabra University)
Mataró
Spain

Antonietta M. Esposito
Istituto Nazionale di Geofisica e
  Vulcanologia, sezione di Napoli
  Osservatorio Vesuviano
Naples
Italy

Gennaro Cordasco
Department of Psychology
Seconda Universita di Napoli and IIASS
Caserta
Italy

Thomas Drugman
University of Mons, TCTS Lab.31,
  Boulevard Dolez
Mons
Belgium

Jordi Solé-Casals
Data and Signal Processing Research Group
University of Vic
Vic
Spain

Francesco Carlo Morabito
Università degli Studi Mediterranea di
  Reggio Calabria
Reggio Calabria
Italy

ISSN 2190-3018   ISSN 2190-3026   (electronic)
Smart Innovation, Systems and Technologies
ISBN 978-3-319-28107-0   ISBN 978-3-319-28109-4   (eBook)
DOI 10.1007/978-3-319-28109-4

Library of Congress Control Number: 2015958848

Printed on acid-free paper

This Springer imprint is published by SpringerNature
The registered company is Springer International Publishing AG Switzerland

Preface

This volume brings together through a peer-revision process advanced research results obtained on nonlinear speech processing, following the tradition initiated by the European COST Action 277: "Nonlinear Speech Processing" (http://www.cost.eu/COST_Actions/ict/277). The research published in this book was discussed for the first time at the 7th edition of the International Workshop on "Nonlinear Speech Processing (NOLISP)" held in Vietri sul Mare, Italy, on May 18–20, 2015.

The workshop afforded a change of perspective in nonlinear speech processing, where the research focus moved from "engineering tools" to "interactional exchanges" and asked for investigations of coding/decoding and computational processes, and social and cognitive speech features for improving the quality of life of end users of communicative interfaces exploiting speech as the main communicative tool for human–machine interaction. The consequences should result in the development of autonomous, adaptive voice user interfaces (VUIs) able to exploit linguistic and paralinguistic information and allow free form of conversations. This approach will foster traverse investigations on the multifunctional role of speech and multimodal communication modes that account for gestures, emotions, and social signal processing for developing friendly and socially believable interactive dialogue systems.

The editors would like to thank the staff of the International Institute for Advanced Scientific Studies (IIASS), who provided precious technical support in the organization of the NOLISP 2015. In addition, the editors are grateful to the contributors for making this book a scientifically stimulating compilation of new

and original ideas, and the NOLISP International Scientific Committee for their
rigorous and invaluable scientific revisions, their dedication, and their priceless
selection process.

November 2015   Anna Esposito
                                                       Marcos Faundez-Zanuy
                                                       Antonietta M. Esposito
                                                         Gennaro Cordasco
                                                          Thomas Drugman
                                                          Jordi Solé-Casals
                                                    Francesco Carlo Morabito

International Scientific Committee

Frédéric Bettens, University of Mons, Belgium
Tony Belpaeme, Plymouth University, UK
Baris Bozkurt, Bahcesehir University, Turkey
Štefan Beňuš, Constantine the Philosopher University, Nitra, Slovakia
Jonas Beskow, Royal Institute of Technology, Sweden
Jean-Francois Bonastre, Universit d'Avignon, France
Nikolaos Bourbakis, Wright State University, Dayton, USA
Joao Cabral, University College Dublin, Ireland
Angelo Cangelosi, Plymouth University, UK
Germán Castellanos, National University of Colombia
Michele Ceccarelli, Universit del Sannio, Italy
Mohamed Chetouani, Universit Pierre et Marie Curie, France
Gérard Chollet, CNRS URA-820, ENST, TSI Department, France
Anton Cizmar, Technical University of Kosice, Slovakia
Gennaro Cordasco, Seconda Università di Napoli and IIASS, Italy
Nicholas Costen, Manchester Metropolitan University, UK
Christophe Dalessandro, LIMSI, France
Vlado Delic, University of Novi Sad, Serbia
Alessandro Di Nuovo, Plymouth University, UK
Thomas Drugman, University of Mons, Belgium
Gilles Degottex, University of Crete, Greece
Stéphane Dupont, University of Mons, Belgium
Daniel Erro, University of the Basque Country, Bilbao, Spain
Anna Esposito, Seconda Universit di Napoli and IIASS, Italy
Antonietta Maria Esposito, Osservatorio Vesuviano, Napoli, Italy
Marcos Faúndez-Zanuy, ESUPT, Barcelona, Spain
Milan Gnjatovič, University of Novi Sad, Serbia
Juan Ignacio Godino, UPM, Spain
Pedro Gómez-Vilda, UPM, Spain
Guillaume Gravier, CNRS, France
Francis Grenez, Free University of Bruxelles, Belgium

Jonathan Harrington, Ludwig-Maximilians University of Munich, Germany
Amir Hussain, University of Stirling, UK
Jozef Juhár, Technical University Kosice, Slovak Republic
John Kane, Trinity College Dublin, Ireland
Maria Koutsombogera, Institute for Language and Speech Processing, Greece
Bernd Kröger, Aachen University, Germany
Gernot Kubin, Graz University of Technology, Austria
Karmele Lopez-De-Ipina, Basque Country University, Spain
Jorge Lucero, University of Brasilia, Brazil
Saturnino Luz, University of Edinburgh, UK
Rytis Maskeliunas, Kaunas University of Technology, Lithuania
Jiří Mekyska, Brno University of Technology, Czech Republic
Francesco C. Morabito, Universit "Mediterranea" di Reggio Calabria, Italy
Rieks Op Den Akker, University of Twente, The Netherlands
Eros Pasero, Politecnico di Torino, Italy
Francesco Piazza, Universit Politecnica delle Marche, Italy
Jiří Přibil, Academy of Sciences, Czech Republic
Anna Přibilová, Slovak University of Technology, Slovakia
Steve Renals, University of Edinburgh, Scotland
Korin Richmond, University of Edinburgh, Scotland
Matej Rojc, University of Maribor, Slovenia
Jean Rouat, University of Sherbrooke, Canada
Zsófia Ruttkay, Moholy-Nagy University of Art and Design, Budapest
Giampiero Salvi, Royal Institute of Technology, Stockholm, Sweden
Michele Scarpiniti, Universit di Roma "La Sapienza", Italy
Jean Schoentgen, Free University of Bruxelles, Belgium
Bjorn Schuller, Munich University of Technology, Germany
Zdenek Smékal, Brno University of Technology, Czech Republic
Simone Scardapane, Universit di Roma "La Sapienza", Italy
Michele Scarpiniti, Universit di Roma "La Sapienza", Italy
Jordi Solé-Casals, University of Vic, Spain
Stefano Squartini, Universit Politecnica delle Marche, Italy
Antonio Satue-Villar, EUPMT, Barcelona, Spain
Björn W. Schuller, Imperial College, UK and University of Passau, Germany
Yannis Stylianou, University of Cambridge, England
Jing Su, Trinity College Dublin, Ireland
Jianhua Tao, Chinese Academy of Sciences, P.R. China
Isabel Trancoso, INESC, Portugal
Carlos M. Travieso-Gonzalez, University of Las Palmas de GC, Spain
Alda Troncone, Second University of Naples, Italy
Christophe Veaux, University of Edinburgh, Scotland
Alessandro Vinciarelli, University of Glasgow, UK
Fernando Villavicencio, Yamaha, Japan
Hannes Högni Vilhjlmsson, Reykjavik University, Iceland
Carl Vogel, Trinity College of Dublin, Ireland

Junichi Yamagishi, University of Edinburgh, Scotland
Aurelio Uncini, Universit di Roma "La Sapienza", Italy
Jerneja Žganec Gros, Alpineon, Development and Research, Slovenia

Sponsoring Institutions

Seconda Università degli Studi di Napoli
International Speech Communication Association (ISCA)
International Institute for Advanced Scientific Studies "E.R. Caianiello" (IIASS)
Società Italiana Reti Neuroniche (SIREN)

Contents

Part I
Nonlinear Speech Processing: An Introduction

A Decade of Encouraging Speech Processing "Outside of the Box"—A Foreword

Björn W. Schuller

Abstract A perspective on this year's Non-Linear Speech Processing effort is given focussing on alternative approaches. In fact, it was marked by a strong presence of the "younger" field of paralinguistic speech analysis. Beyond the more mature recognition of emotion, a significant interest in health and care applications can be noticed. This includes Alzheimer's disease, cognitive load level, depressive disorders, and Parkinson's disease, and reaches to language impairment, general pathology as well as speech interfaces for the elderly. In fact, even speaker identification has seen a study on the influence of emotionally converted speech. Related to speech analysis, the front-end side along the usual chain of speech processing was strongly represented. Backend machine learning includes extreme learning machines, unsupervised clustering, and deep representations.

The chapters of this collection are an extension of the work the authors had presented at the 2015 seventh biennial international Non-linear Speech Processing (NOLISP) workshop. NOLSIP 2015 has—as ever in this series—invited *alternative* approaches in speech processing. This is a more than crucial focus considering the growing "age" of computational speech processing—in particular speech and speaker recognition benefit from more than six and roughly five decades of experience by now. This may appear young in comparison to many traditional fields of research, but relating it to the birth hour of Computer Science and according milestones beyond Boole's algebra in the later 1840s such as the Turing machine in the later 1930s, makes Speech Processing increasingly a rather "traditional" area of research in Computer Science. While these decades have overall been marked by significant progress, one could also repeatedly witness "plateaus of stagnation" such as last in the noughties. It has been exactly the thinking *offside the "conventional"* that continually helped to overcome the pitfall of overly trusting in one direction. A good example of such

B.W. Schuller (✉)
Machine Learning Group, Imperial College London, London, UK
e-mail: bjoern.schuller@imperial.ac.uk

B.W. Schuller
Chair of Complex and Intelligent Systems, University of Passau, Passau, Germany

© Springer International Publishing Switzerland 2016
A. Esposito et al. (eds.), *Recent Advances in Nonlinear Speech Processing*,
Smart Innovation, Systems and Technologies 48,
DOI 10.1007/978-3-319-28109-4_1

3

thinking "outside of the box" was the recent change of emphasis from statistical to neural approaches within the "deep learning wave".

In fact, this collection of extended contributions of the NOLISP 2015 program is marked by a strong presence of the "younger" roughly two decades aged field of paralinguistic speech analysis. Beyond the "more mature" recognition of emotion (Camilo Vasquez et al.), a significant part is dedicated to health and care applications. This includes chapters on speech analysis related to Alzheimer's disease (López-De-Ipiña et al.), cognitive load level (Luz and Su), depressive disorders (Esposito et al. and Kiss et al.), and Parkinson's disease (Gómez-Vilda et al. and Mekyska et al.); the content further reaches to new findings regarding specific language impairment (Dacewicz, Nowak, and Szelag), general pathology (Szelag and Dacewicz) as well as speech interfaces for the elderly (Wang et al.). Further *paralinguistic* speaker classification contained deals with speaker age classification (Muñoz-Mulas et al.), and personality traits, role, and conflict in a summary of the keynote that had been given by Alessandro Vinciarelli. Similarly concerned with social aspects and speaking style is the summary of the other keynote that had been presented by Jonathan Harrington; it stresses the relevance of contextual aspects and sound change. In fact, even speaker identification featured in this collection deals with the influence of *emotionally* converted speech (Pribil and Pribilova) besides speaker identification in "more usual" settings (Ghezaiel, Ben Slimane, and Ben Braiek).

Related to speech analysis, the *front-end* side along the usual chain of speech processing is strongly represented: Prosody (Babu and Sao), and in particular pitch estimation (Jlassi, Bouzid, and Ellouze) as well as glottal closure (Smidi, Bouzid, and Ellouze) and glottal features (Lazaro-Carrascosa and Gómez-Vilda), modelling of diphthong articulation (Carmona-Duarte et al.), and stationary voiced speech (Zammel and Ellouze) make novel speech features and modelling a further major topic of the book. The pre-processing and enhancement usually preceding this modelling along this chain of processing is represented in this volume by new methods for nonlinear acoustic echo cancellation (Comminiello et al.).

NOLISP was this year jointly organised with the 25th Italian Workshop on Neural Networks (WIRN 2015)—a perfect match in these days of "neural renaissance". It thus seems not surprising that more *"backend"* machine learning-focused contributions could also be found in its program. The extended original contributions in this collection reach from experience with extreme learning machines (Della Porta et al.), and unsupervised clustering and deep representations (Salvi) to "non-neural network-related" chapters, e.g., concerned with Markov chains (Singh).

Finally, on the "opposite end" of analysis, two contributions consider alternative approaches for hidden Markov models (Sulír and Juhár), and deep neural networks (Takaki and Yamagishi) for speech synthesis.

Overall, NOLISP itself celebrated its first decade's anniversary after the first edition held in 2005 in this year. Unaffected by this, it faithfully kept *following* its aim to provide space to those works *not following* the "mainstream". Besides alternative approaches, the still "somewhat alternative" Computational Paralinguistics made up for the lion's share of its program in 2015. It will be exciting to see which of this year's extended contributions featured in this collection will contribute to shape the future of non-linear speech processing.

Recent Advances in Nonlinear Speech Processing: Directions and Challenges

Anna Esposito, Marcos Faundez-Zanuy, Antonietta M. Esposito,
Gennaro Cordasco, Thomas Drugman, Jordi Solé-Casals and
Francesco Carlo Morabito

Abstract Humans have very high requirements and expectations when communicating through speech, other than simplicity, flexibility and easiness of interaction. This is because voice interactions do not require cognitive efforts, attention, and memory resources. Voice technologies are however still constrained to use cases and scenarios giving the existing limitations of speech synthesis and recognition systems. Which is the status of nonlinear speech processing techniques and the steps made for cross-fertilization among disciplines? This chapter will provide a short overview trying to answer the above question.

Keywords Nonlinear speech processing · Socially believable voice user interfaces · Sound changes · Social and emotional speech features

A. Esposito (✉)
Department of Psychology, Seconda Università di Napoli and IIASS, Caserta, Italy
e-mail: iiass.annaesp@tin.it

M. Faundez-Zanuy
Escola Superior Politècnica Tecnocampus (Pompeu Fabra University), Mataró, Spain
e-mail: faundez@tecnocampus.cat

A.M. Esposito
Istituto Nazionale di Geofisica e Vulcanologia, sezione di Napoli
Osservatorio Vesuviano, Rome, Italy
e-mail: antonietta.esposito@ingv.it

G. Cordasco
Department of Psychology, Seconda Università di Napoli and IIASS, Caserta, Italy
e-mail: gennaro.cordasc@unina2.it

T. Drugman
University of Mons, TCTS Lab.31, Boulevard Dolez, Mons, Belgium
e-mail: Thomas.DRUGMAN@umons.ac.be

J. Solé-Casals
Data and Signal Processing Research Group, University of Vic, Barcelona, Spain
e-mail: jordi.sole@uvic.cat

F.C. Morabito
Università degli Studi "Mediterranea" di Reggio Calabria, Reggio Calabria, Italy
e-mail: morabito@unirc.it

© Springer International Publishing Switzerland 2016
A. Esposito et al. (eds.), *Recent Advances in Nonlinear Speech Processing*,
Smart Innovation, Systems and Technologies 48,
DOI 10.1007/978-3-319-28109-4_2

1 Introduction

Even though contextual instances play a fundamental role in delineating the most appropriate communication tools for implementing successful interactional exchanges [12], nevertheless, spoken messages remain naturally preferred and extremely effective among humans. This is substantiated by the fact that speech based information communication technologies (ICT) are largely accepted and favored among persons. To our knowledge, visual telecommunication tools, such as teleconferencing, are still at an early stage of acceptance, because their *"perceived ease of use* (PEOU)", and *"perceived usefulness* (PU)", are strongly affected by both *"individual factors such as anxiety and self-efficacy, and institutional factors such as institutional support and voluntariness"* [26, p.118]. On the contrary, Voice User Interfaces (VUIs), had proven to be largely accepted to the extent that 65+ aged elders are enthusiast to be assisted and monitored for their chronic diseases by a static speaking face [8].

A spoken message produces a precise physical object, a wave of sounds, through which an individual communicates ideas and beliefs, shares knowledge, express needs, feelings, and emotions. The everyday simplicity and flexibility of a such acoustic event in serving as a "container" of countless superimposing and interweaving information, is impressive. The elementary "wave of sounds" will take on several encoding channels, where different streams of data flow together to efficiently build up and successfully shape human exchanges. Among all these encodings, the linguistic code is undoubtedly the most important. It exploits a predefined and shared communication protocol (the language[1]) that allows interactants to decipher a substantial part of the semantic meaning of the delivered message. However, there is a lot of additional information normally sent through speech. Psycholinguistic studies have shown that meanings are conveyed not only by words (intended here as lexicon). During speech production, there exist multiple sets of non-lexical expressions carrying on specific communicative values. Typical non-lexical communicative events at the paralinguistic speech level are, for example, empty and filled pauses signaling, among many other functions, mood states; vocalizations signaling positive or negative feedbacks (*"aah"*, *"hum"*); speech repairs signaling speakers cognitive and emotional states, as well as discourse planning/re-planning strategies; and intonational phrases contour changes allowing to disambiguate meanings [6, 7, 10, 12–14]. The abovementioned speech resources are powerful enough to fulfill plenty of communicative needs without the intervention and independently from the linguistic code, since the process of encoding/decoding for this information is very likely affected by cultural, unconscious, and instinctive communication mechanisms rather than by language production/comprehension rules.

In addition, it is well known that communicative exchanges among humans are not achieved only through speech and linguistic vocal expressions. Written and visual

---

[1] Here "language" is intended to be "the verbal language" as opposed to other general meanings of the term. The interpretation of a "language" as a code can be found in De Saussure [9].

channels, convey linguistic and paralinguistic information that complement or substitute spoken messages and gestures achieve the same pragmatic and semantic speech function [12, 18]. However, at the current technological stage there are few ICT technologies exploiting these channels: speech technologies predominate among all of them and are favorite with respect to visual, graphical and text interfaces. The ultimate speech ICT objectives are guided by the willingness to improve voice services in telecommunication systems, providing a high quality speech synthesis, more efficient speech coding, effective speech recognition, speaker identification, and speaker verification systems in order to significantly spread the VUIs acceptance for information systems such as the mobile Internet (by improving speech synthesis and recognition) and the future generations of wireless communication networks (by improving speech coding).

2 Beyond Nonlinear Speech Processing

The nonlinear approach to speech processing had produced advances in several speech engineering fields such as coding, transmission, compression, and synthesis among others, as well as, advances beyond the engineering approach. This is because the functional role of speech, being a human ability, is not constrained to a finite scope and therefore, investigations in one field had produced results in another. Among the topics that had exploited for long time and still exploit nonlinear techniques, it is worth to mention Speech Coding, intended as the ability of an algorithm to code speech in a compact bit-stream such that the amount of transmitted data (the bit rate) would be as low as possible to accommodate transmission channel constraints while preserving speech intelligibility and pleasantness [1, 2, 20]. Low-rate speech coding algorithms have been developed for interactive multimedia services on packet-switched networks such as mobile radio networks, Internet, and mobile network user base, and even more very low bit rate coding at consumer quality will be demanded by the future ICT systems [21, 22, 31].

Two topics of highly nonlinear relevance are Speech Synthesis and Recognition. Humans have very high requirements and expectations when dealing with VUIs, other than simplicity, flexibility and easiness of interaction. This is because voice interactions are an ordinary tool of exchanges among them and do not require, on the user side, cognitive efforts, attention, and memory resources as in the case of graphical and text interfaces. Voice exchanges between humans and machines eliminate delays caused by option menus and can provide very rapidly and complex verbal responses. However current VUIs are not free of constraints. VUIs represent a complex interface option for systems developers since the underlying automated speech recognition (ASR) and Text to Speech (TTS) technology is constrained to context based and speaker dependent applications. Free-form of human-machine conversations are not provided by the current speech technologies. Improvements in dialog management resources are still addressed to specific use scenarios varying from allowing health users to surf the World Wide Web to more complex applications such monitoring the wellbeing of elderly people, which add, to the complexity of the free-form of conversations also those related to poor speech production (and then more complex efforts for its recognition) because of possible fine motor articulatory impairments due to the age [8, 24, 25, 30]. Current commercial voice enabled systems are Webtalk (http://www.pcworld.com/article/98603/article.html) developed by Microsoft, and Siri (http://www.apple.com/ios/siri/) developed by Apple. These systems are not free of criticisms and still constrained in the dialogue management to be speaker-dependent, with a restricted dictionary, and favorable environmental conditions. These limitations are mostly due to the many sources of variability affecting the speech signals coarsely grouped by Esposito [16] as: "a) phonetic variability (i.e. the acoustic realizations of phonemes are highly dependent on the context in which they appear), b) within-speaker variability (as result of changes in the speakers physical and emotional state, speaking rate, voice quality), c) across-speaker variability (due to differences in the socio-linguistic background, gender, dialect, and size and shape of the vocal tract), and d) acoustic variability (as result of changes in the environment as well as the position and the characteristics of the transducer)". Reliable and effective speech recognition and synthesis applications must be able to handle efficiently these variabilities knowing at any stage of the speech recognition/synthesis process which source more than the others is affecting the system efficiency and performance. The general assumption behind these investigations is "that there are rules governing speech variability and such rules can be learned and applied in practical situations" [15, 16]. This point of view is not generally accepted (see [23] for an alternative point of view), since it is related to the classical problem of reconciling the physical and linguistic description of speech, i.e. the invariance issue. Five decades of research in nonlinear speech processing seems to bring convincing arguments on the role of the context (the cultural, organizational, and physical context) in the human communications [12] suggesting to consider the invariance issue context dependent to a certain extent. Two more nonlinear engineering topics such as Voice Analysis, and Conversion (where the quality of the human voice is analysed for clinical and phonetics applications and where techniques for the manipulation of voice characters) produced the flourishment of new speech research fields and new speech applications, such as the analysis of emotional vocal expressions in order to identify speech acoustic emotional features and be able to detect emotional states from speech [3–5, 17, 27, 28] and even more psychopathological disorders such as depression, stress and anxiety [11, 19, 29].

The nonlinear approach to speech processing had gone beyond the acoustic and engineering approach to speech processing, extending its research to the psychological, social, and organizational implications derived from exchanges that are not anymore only among humans, being an automatic system involved. However, in order to be an efficient and effective exchange, the richness of the speech signal must be preserved combining appropriately technological constraints and its social and functional role.

3　Contents of this Book

It took over 50 years to realize that speech is beyond speech and therefore nonlinear speech processing should go beyond nonlinear techniques and exploits heuristic and psychological models of human interaction in order to succeed in the implementations of socially believable VUIs and applications for human health and psychological support. This book is signaling advances in these directions taking into account the multifunctional role of speech and what is "outside of the box" (see Björn Schuller's foreword). To this aim, the book is organized in 6 sections, each collecting a small number of short chapters reporting advances "inside" and "outside" themes related to nonlinear speech research. The themes emphasize theoretical and practical issues for modelling socially believable speech interfaces, ranging from efforts to capture the nature of sound changes in linguistic contexts and the timing nature of speech; labors to identify and detect speech features that help in the diagnosis of psychological and neuronal disease, attempts to improve the effectiveness and performance of Voice User Interfaces, new front-end algorithms for the coding/decoding of effective and computationally efficient acoustic and linguistic speech representations, as well as investigations capturing the social nature of speech in signaling personality traits, emotions and improving human machine interactions.

The coarsely arrangement in 6 scientific sections should be considered only a thematic classification. The sections are closely connected and provide fundamental insights for the cross-fertilization of different disciplines. All the chapters collected in each section are original and never published before. In addition, all the chapters benefited from the live interactions in person among the participants of the successful meeting in Vietri sul Mare under the egide of the 7th biennial international workshop on Non-Linear Speech Processing (NOLISP 2015) which had initiated alternative approaches to speech processing according to the research tradition proposed by the COST Action 277 (http://www.cost.eu/COST_Actions/ict/277).

4　Conclusions

The readers of this book will get a taste of the major research areas on nonlinear speech processing, different visions on the multifunctional role of speech, different methodologies for analyzing and detecting important speech features, psychological, social, and cognitive disease, and how nonlinear speech processing interact with cognitive and social processes and can shed light on their comprehension and understanding. The research topics proposed by the book are particularly computer science, engineering, signal processing and human-computer interaction oriented and the contributors to this volume are leading authorities in their respective fields. However, interesting psychological, and cognitive aspects are also captured and discussed, letting the book to go, as speech itself, beyond and across scientific disciplines.

References

1. Arjona Ramírez, M., Minami, M.: Technology and standards for low-bit-rate vocoding methods. In: Bidgoli, H. (ed.) The Handbook of Computer Networks, vol. 2, pp. 447–467. Wiley, New York (2011)
2. Arjona Ramírez, M., Minami, M.: Low bit rate speech coding. In: Proakis, J.G. (ed.) Wiley Encyclopedia of Telecommunications, vol. 3, pp. 1299–1308. Wiley, New York (2003)
3. Atassi, H., Esposito, A., Smekal, Z.: Analysis of high-level features for vocal emotion recognition. In: Proceedings of 34th IEEE International Conference on Telecommunication and Signal Processing (TSP), pp. 361–366 (2011)
4. Atassi, H., Riviello, M.T., Smekal, Z., Hussain, A., Esposito, A.: Emotional vocal expressions recognition using the cost 2102 italian database of emotional speech. In: Esposito, A., et al. (eds.) Development of Multimodal Interfaces: Active Listening and Synchrony, LNCS 5967, pp. 255–267. Springer, Berlin, Heidelberg (2010)
5. Atassi, H., Esposito, A.: Speaker independent approach to the classification of emotional vocal expressions. In: Proceedings of IEEE Conference on Tools with Artificial Intelligence (ICTAI 2008), vol. 1, pp. 487–494 (2008)
6. Butterworth, B.L., Beattie, G.W.: Gestures and silence as indicator of planning in speech. In: Smith, P.T., Campbell, R.N. (eds.) Recent Advances in the Psychology of Language, pp. 347–360. Olenum Press, New York (1978)
7. Chafe, W.L.: Cognitive constraint on information flow. In: Tomlin, R. (ed.) Coherence and Grounding in Discourse, pp. 20–51. John Benjamins, Amsterdam (1987)
8. Cordasco, G., Esposito, M., Masucci, F., Riviello, M.T., Esposito, A., Chollet, G., Schlögl, S., Milhorat, P., Pelosi, G.: Assessing voice user interfaces: the vAssist system prototype. In: 5th IEEE International Conference on Cognitive InfoCommunications, pp. 91–96. Vietri sul Mare, 5–7 Nov 2014
9. De Saussure, F.: Cours de linguistique générale. Editions Payot, Paris (1922)
10. Esposito, A., Esposito, A.M., Vogel, C.: Needs and challenges in human computer interaction for processing social emotional information. Pattern Recogn. Lett. **66**, 41–51 (2015)
11. Esposito, A., Esposito, A.M., Likforman, L., Maldonato, M.N., Vinciarelli, A.: On the significance of speech pauses in depressive disorders: results on read and spontaneous narratives. In this volume (2015)
12. Esposito, A.: The situated multimodal facets of human communication. In: Rojc, M., Campbell, N. (eds.) Coverbal Synchrony in Human-Machine Interaction, ch. 7, pp. 173–202. CRC Press, Taylor & Francis Group, Boca Raton, FL (2013)
13. Esposito, A., Marinaro, M.: What pauses can tell us about speech and gesture partnership. In: Esposito, A., et al. (eds.) Fundamentals of Verbal and Nonverbal Communication and the Biometric Issue. NATO Publishing Series, vol. 18, pp. 45–57. IOS Press, The Netherlands (2007)
14. Esposito, A., Bourbakis, N.G.: The role of timing in speech perception and speech production processes and its effects on language impaired individuals. In: Proceedings of the 6th International IEEE Symposium on BioInformatics and BioEngineering (BIBE), pp. 348–356 (2006)
15. Esposito, A.: The importance of data for training intelligent devices. In: Apolloni, B., Kurfess, C. (eds.) From Synapses to Rules: Discovering Symbolic Knowledge from Neural Processed Data, pp. 229–250. Kluwer Academic Press, Dordrecht (2002)
16. Esposito, A.: Approaching speech signal problems: an unifying viewpoint for the speech recognition process. In: Suarez Garcia, S., Baron Fernandez, R. (eds.) Memoria of Taller Internacional de Tratamiento del Habla, Procesamiento de Vos y el Language, CIC-IPN Obra Compleata (2000). ISBN: 970-18-4936-1
17. Galanis, D., Karabetsos, S., Koutsombogera, M., Papageorgiou, H., Esposito, A., Riviello, M.T.: Classification of emotional speech units in call centre interactions. In: Proceedings of 4th IEEE International Conference on Cognitive Infocommunications (CogInfoCom2013), pp. 403–406. Budapest, Hungary, 2–5 Dec 2013

18. Kendon, A.: Gesture: Visible Action as Utterance. Cambridge University Press, Cambridge (2004)
19. Kiss, G., Tulics, M.G., Sztahó, D., Esposito, A., Vicsi, K.: Language independent detection possibilities of depression by speech. In this volume (2015)
20. Kroon, P.: Evaluation of speech coders. In: Paliwal, K.K., Bastiaan Kleijn, W. (eds.) Speech Coding and Synthesis, pp. 467–494. Elsevier Science, Amsterdam (1995)
21. Gibson, J.D.: Speech coding methods, standards, and applications. IEEE Circuits Syst. Mag. **5**(4), 30–49 (2005)
22. Faundez-Zanuy, M., Janer, L., Esposito, A., Satue-Villar, A., Roure, J., Espinosa-Duro, V. (eds.): Nonlinear Analyses and Algorithms for Speech Processing, LNAI 3817. Springer, Berlin, Heidelberg (2006)
23. Lindblom, B.: Explaining phonetic variation: a sketch of the H&H theory. In: Hardcastle, W., Marchal, A. (eds.) Speech Production and Speech Modeling, pp. 403–439. Kluwer, Dordrecht (1990)
24. Meena, R., Skantze, G., Gustafson, J.: Data-driven models for timing feedback responses in a map task dialogue system. Comput. Speech Lang. **28**, 903–922 (2014)
25. Milhorat, P., Schlögl, S., Chollet, G., Boudyy, J., Esposito, A., Pelosi, G.: Building the next generation of personal digital assistants. In: Proceedings of 1st IEEE International Conference on Advanced Technologies for Signal and Image Processing–ATSIP'2014, pp. 458–463. Sousse, Tunisia, 17–19 Mar 2014. ISSN 978-1-4799-4888-8/14/
26. Park, N., Rhoads, M., Hou, J., Lee, K.M.: Understanding the acceptance of teleconferencing systems among employees: an extension of the technology acceptance model. Comput. Hum. Behav. **39**, 118–127 (2014)
27. Ringeval, F., Eyben, F., Kroupi, E., Yuce, A., Thiran, J.P., Ebrahimi, T., Lalanne, D., Schuller, B.: Prediction of asynchronous dimensional emotion ratings from audiovisual and physiological data. Pattern Recogn. Lett. Elsevier (2014)
28. Schullerm, B.: Deep learning our everyday emotions: a short overview. In: Bassis et al. (eds.) Advances in Neural Networks: Computational and Theoretical Issues. Series: SIST Series, vol. 37, pp. 339–346. Springer, Berlin, Heidelberg (2015)
29. Scherer, S., Stratou, G., Lucas, G., Mahmoud, M., Boberg, J., Gratch, J., Rizzo, A., Morency, L.P.: Automatic audio-visual behaviour descriptors for psychological disorder analysis. Special Issue on Best of Face and Gesture 2013: Image Vis. Comput. 32(10), 648–658 (2014)
30. Skantze, G., Hjalmarsson, A.: Towards incremental speech generation in conversational systems. Comput. Speech Lang. **27**, 243–262 (2013)
31. Stylianou, Y., Faundez-Zanuy, M., Esposito, A. (eds.): Progress in Nonlinear Speech Processing, LNCS 4391. Springer, Berlin, Heidelberg (2007)

Part II
Features of Sound Change

The Relationship Between the (Mis)-Parsing of Coarticulation in Perception and Sound Change: Evidence from Dissimilation and Language Acquisition

Jonathan Harrington, Felicitas Kleber and Mary Stevens

Abstract The study is concerned with whether historical sound change is more likely to occur when coarticulation, or the way that speech sounds overlap with and influence each other in time, is misaligned in production and perception. The focus of the first experiment was on long-range coarticulatory lip-rounding that has been linked with historical dissimilation. A perception experiment based on present-day Italian showed that inherently lip-rounded segments were more likely to be masked— and thereby erroneously deleted—in hypoarticulated speech. The second experiment tested whether the mismatch between the modalities was more likely in young children than in adults. For this purpose, first language German speakers participated in a forced-choice perception experiment in which they categorised German back and front vowels in coarticulatory non-fronting and fronting consonantal contexts. Children's ability to normalise for coarticulation was shown to be less than that of the adults. Taken together, the results suggest that sound change can occur when coarticulatory relationships are perceptually obscured due to a hypoarticulated speaking style causing consonants to be camouflaged in the case of dissimilation and variants to approximate those that are strongly influenced by coarticulation in the case of diachronic back vowel fronting.

1 Introduction

Research in the last 30–40 years has shown a relationship between contextual variation in speech communication and historical change. A well-known example is synchronic transconsonantal vowel coarticulation [24, 52] that has led to the sound

J. Harrington (✉) · F. Kleber · M. Stevens
Institute of Phonetics and Speech Processing, Ludwig-Maximilians
University of Munich, Munich, Germany
e-mail: jmh@phonetik.uni-muenchen.de

F. Kleber
e-mail: kleber@phonetik.uni-muenchen.de

M. Stevens
e-mail: mes@phonetik.uni-muenchen.de

© Springer International Publishing Switzerland 2016   15
A. Esposito et al. (eds.), *Recent Advances in Nonlinear Speech Processing*,
Smart Innovation, Systems and Technologies 48,
DOI 10.1007/978-3-319-28109-4_3

change by which umlaut has developed in some languages (e.g. present-day German *Füße* /fyse/ and present-day English 'feet' from Proto-Germanic /fotiz/). The general aim in this chapter is to consider the mechanisms by which diachronic change can take hold, given what is known about the dynamics of speech production and their relationship to perception. The focus will be on the types of sound change that have been documented in numerous languages and whose bases are in coarticulation: that is in how speech sounds overlap with and influence each other in time.

Many physiological, acoustic, and perceptual studies are consistent with the idea developed from action theory [15] through to articulatory phonology [7] that speech production can be modelled as the orchestration of autonomous gestures that wax and wane in time [45] so that, within any given time window multiple sounds make contributions in different degrees of strength to the acoustic signal [59]. Thus, in producing *queen* /kwin/, the tongue-dorsum raising of /i/ is likely to overlap partially or entirely both with the preceding lip-rounding from /w/ and with a lowered velum in anticipation of the following /n/. This simultaneous production, coarticulation or coproduction of multiple gestures from successive speech sounds can also easily extend across major prosodic boundaries, especially in the case of liquids [23, 32, 61]. The important point as far as modelling sound change from coarticulation is concerned is that, to use an apt metaphor from Lindblom [43], speech is 'big band' in which the gestures of speech production are independently controlled and each make their own contribution to the acoustic signal, analogously to the acoustic contribution towards realising a common musical arrangement that is made by the independently controlled and separate instruments of an orchestra.

As far as the listener is concerned, numerous experiments suggest that coarticulation is perceived analogously to its production. The evidence for this derives from experiments showing how an identical acoustical signal is differently perceived depending on the context in which it is embedded [21, 46]. For example, many listeners hear a nasal vowel as *oral* when it is surrounded by nasal consonants [34]. An explanation for this finding is that listeners parse the acoustic signal into the overlapping articulatory gestures that could have given rise to it [17, 20]. For this example, listeners perceive the temporally overlapping tongue dorsum movements for the vowel and lowered velum of the nasal as autonomously coproduced gestures. A consequence of perceived coproduction is that nasalisation in perception is associated not with the vowel but with the nasal consonant that caused it: that is, listeners factor nasalisation from the acoustic signal of the vowel [4, 5]. It is in this sense that some have argued for parity or a common currency between coarticulation in production and coarticulation in perception [16]. For some researchers, there must necessarily be parity because gestures are directly perceived in the speech signal [19]. In the theory to be developed in the present chapter, such parity across production and perception in processing coarticulation is considered to be the condition that obtains only under stability, i.e. when no sound change is taking place. Moreover, we propose that sound change can occur under the perhaps rarer condition in which the production and perception of coarticulation are out of alignment: that is, when listeners perceive or parse coarticulation in a way that is different from its production.

This theory is closely informed by Ohala's [48–50] model of sound change and extensions thereof (e.g. [41, 57]) in which occasional ambiguities in the transmission of coarticulation from a speaker to hearer can be a source of sound change. A well-known example is the epenthetic stop insertion that synchronically gives rise to variations such as /drɛmt, drɛmpt/ ('dreamed') that are related to sound changes such as *empty* from old English /æmtiɡ/. Where then does the /p/ come from? A likely answer is that a /p/ can be perceived if the lip closure for /m/ is released not synchronously but after the oral closure leading to a bilabial stop or doubly articulated [p͡t]. Notice that the listener must have heard a /p/ even though no such unit formed part of speech production. That this must be so can be seen in the derivation of names such as 'Hampton' which arose by combining in Old English the surname 'Ham' (and importantly not 'Hamp') with 'tun'. Thus, the part of the signal corresponding to the overlapping lip-constriction and /t/ closure that must have occurred in /æmtiɡ/ and /hamtun/ has been decontextualised by the listener, because it is not interpreted in relation to the phonetic context that gave rise to it. This part of the signal has (with the passage of time) instead been *phonologised* because it has come to be permanently associated (in 'empty' and 'Hampton') with a /p/ phoneme where none had originally existed.

Speech communication also varies substantially in speaking style. This can be in a social sense, as when speakers adapt their style to take account of the social status of the interlocutor (e.g. [30]). Moreover, there is, of course, well-documented evidence of an association between the adaptation of speaking style towards a more prestigious social class and sound change [38]. Here we shall be concerned not with social variation, but instead with the adaptation in speech production depending on the extent to which the meaning of the signal is predictable from context. According to Lindblom [42], speech is produced with a high degree of clarity or *hyper*articulation when the listener has to rely almost entirely on the signal to understand it. This might happen in introducing a person for the first time, given that there is unlikely to be any prior context or knowledge by which the listener can infer the person's name from context. Local hyperarticulation is likely to occur at points in the signal that are particularly important for understanding what is being said [11]; in stress-accent languages, these points of information focus also typically occur in nuclear accented words [10]. By contrast, a speaker tends to *hypo*articulate the parts of the speech signal in which the listener is predicted to be able to bring to bear contextual knowledge in the broadest sense—sometimes because of a topic that is current in a dialogue, sometimes by means of the knowledge that is assumed to be shared by the speaker and listener [54]. According to Lindblom [42], listeners tend not to process the details of the signal in hypoarticulated speech: firstly, they might not need to because the signal should be highly predictable using top-down information; secondly, hypoarticulated signals may in any case be of less use for decoding meaning if the phonetic content is degraded—such as when vowels are reduced and consonants are strongly lenited, as is typical of a hypoarticulated speaking style. In Lindblom [44], it is when listeners exceptionally process the fine phonetic detail in hypoarticulated speech that a new pronunciation for a word can be added to the lexicon.

Harrington et al. [28, 29] tested whether the types of ambiguities in the transmission of coarticulation—that Ohala considers to be responsible for many kinds of sound change—may be more acute in hypoarticulated speech. They assessed whether prosodic weakening influenced the extent to which listeners adjusted their perceptions for a coarticulatory effect (vowel fronting in VCV coarticulation and polysyllabic shortening). Their results suggested less perceptual adjustment for coarticulation in lexically weak than strong syllables [28] and less adjustment in prosodically deaccented than accented words [29]. Taken together, these results provide some evidence that listeners' phonological categorisations are less influenced by coarticulation in hypoarticulated speech (of which lexically weak syllables and deaccented words are two examples). In this paper, we extend this idea to test whether there is a connection between dissimilation sound changes and the degradation of perceived coarticulation in hypoarticulated speech (see [2] and [6] for a review and analysis of dissimilation in different languages). Dissimilation is a very different type of sound change compared with those that formed the basis of the analysis in Harrington et al. [28, 29] in which the sound changes associated with phonetic variation come about because listeners are presumed to adjust their perceptions insufficiently for coarticulation. Dissimilation, by contrast, comes about according to Ohala [53] because listeners adjust their perceptions *too much* for a presumed coarticulatory effect. So far, there have been very few attempts to reconstruct in the laboratory the synchronic conditions that could lead to dissimilation and the few that have been conducted (e.g. [1]) have found little evidence to support the idea that dissimilation is associated with an over-compensation for coarticulation, as suggested by Ohala [48, 49].

In the last 10 years, various studies have shown that listeners even of the same dialect do not always agree on how to process coarticulation. For example, studies by Beddor [4, 5] have shown that American English listeners vary in how nasalised they perceive a vowel to be before a nasal consonant. Moreover, Yu [63] and Yu et al. [64] have shown how listeners' perception of (and normalisation for) coarticulation is influenced by their personality and social profiles. Such results suggest another potential source of sound change: that individuals or perhaps groups of individuals differ in how a given speech signal is parsed perceptually (see also [33]). Group differences in processing coarticulation perceptually were found for older versus younger subjects for a sound change in progress in Standard Southern British [27, 36].

In this chapter, we consider whether the possibly different ways in which adults and young children perceive coarticulation may be another potential source of sound change. There is of course an extensive literature on the association between sound change and language acquisition [22, 35, 40] with a particular emphasis on demonstrating the commonality between children's misarticulations during acquisition and patterns of sound change. As argued elsewhere [3, 14, 60], there is little evidence for such a direct association and this is also not the type of investigation that is being pursued here. The approach follows instead that of Kleber and Peters [37] who seek to test whether, as less experienced users of the language, children are more likely to have difficulty normalising in perception for coarticulation. Some evidence that this might be so was presented in Nittrouer and Studdert-Kennedy [47] for consonants.

Since that study, there have been no further analyses of whether adults and children process coarticulation differently in perception. Here we extend their analysis for the first time to an investigation of the coarticulatory influences of consonants on vowels (and in another language).

In summary, the aim of this chapter is to test earlier [50] and more recent [4, 36, 41, 57] models of an association between sound change and the perceptual processing of coarticulation. We approach this issue from two very different perspectives. Firstly, by considering how listeners' processing ambiguities could give rise to dissimilation (Sect. 2). Secondly, by analysing whether sound change might arise through the different ways that coarticulation might be processed across two groups of listeners—in this case children and adults (Sect. 3).

2 Sound Change and Dissimilation

The experiment in this section was concerned with the relationship between perceptual processing and a dissimilatory sound change. Dissimilation occurs when one of two similar segments in close proximity changes to become less similar. An example is Grassman's Law under which aspiration disappears when there is another following aspirated stop, e.g. Ancient Greek /t^{h}riks/ 'hair' nominative' but /trik^{h}os/ 'hair' (genitive) derived historically from /t^{h}rik^{h}os/ with initial aspiration. According to Ohala [50], dissimilation can occur when listeners mistakenly attribute part of a speech sound to coarticulation instead of to the speech sound itself. For the example above, sound change comes about because listeners mistakenly interpret the first aspirated segment as being caused by anticipatory coarticulatory spreading of the second /h/. A similar idea is used to explain the sound change whereby the first /w/ was deleted from Latin /kwinkwe/ ('five') leading to /kinkwe/ (and then via a different sound change to /tʃinkwe/ in present-day Italian). The interpretation in Ohala's model is that there is long-range lip-rounding between the two /w/s in /kwinkwe/ that listeners attribute to the second /w/. From another point of view, long-range lip-rounding due to coarticulation occasionally prevents listeners from interpreting the first /w/ as a phonological unit in its own right. Notice that this sound change did not apply to Latin *quindecim* ('fifteen') which is produced with an initial /kw/ in present-day Italian, i.e. /kwinditʃi/). This is because there is no /w/ that occurs later in this word to which coarticulation could be incorrectly attributed.

In the following experiment, two hypotheses were tested. The first was concerned with creating the conditions in the laboratory that could have given rise to dissimilation by testing whether, in present-day Italian, a later occurring /w/ could mask the perception of an initial /w/. Since there are very few words in Italian with a repeated /w/ (*qualunque* being one of the few exceptions), this was done by testing the perception of /kw/ versus /k/ in a target word when followed by a word that did (*quattro*, 'four', /kwat:ro/) or did not (*sette*, 'seven', /set:e/) contain a prevocalic /w/. The second hypothesis was concerned with the effects of hypoarticulation. Here we tested whether the perception of the first /w/ was even more likely to be masked by coarticulation when the target word occurred in a deaccenting/hypoarticulation context. The reasoning behind this follows the arguments of the preceding section that the perceptual parsing of coarticulation may be obscured in hypoarticulated signals: that is, hypoarticulated speech may blur the distinction between lip-rounding due to the presence of an /w/ and the long-range coarticulatory effects of lip-rounding that arise due to the second /w/.

2.1 Method

We created a /kw…k/ continuum and tested the effect of the following word (an initial /kw/ vs. /s/) and prosodic context (accented or deaccented) on listener perception of lip rounding. To create the stimuli, we extracted a single *canto* (/kanto/ 'I sing') token from phrases that had been read aloud and recorded by a female Italian speaker for the purposes of this experiment. We used PSOLA in Praat to synthesize an 11-step continuum from /kanto/ to /kwanto/ (*quanto*, 'how much'). We also lowered F2 in /anto/ to simulate lip rounding throughout the word. We discarded steps 2 and 10 to keep the time taken for the experiment as short as possible for participants. The resulting 9-step *canto…quanto* continuum was inserted into four different carrier phrases that differed according to the following word (*quattro* vs. *sette*) and to whether or not the target word *canto-quanto* was accented (shown in upper case below) or deaccented. In the accented condition, the nuclear accent fell on *canto-quanto* which was synthesised with a large pitch obtrusion appropriate for an L+H* pitch-accent on the first syllable /kan/. In the deaccented condition, the nuclear accent fell on *detto* ('said'), the pitch obtrusion occurred on the first syllable of 'detto' and *canto-quanto* were deaccented (synthesised with a low and flat pitch). The two readings differ in the location of (narrow) focus: thus, the accented condition might be appropriate as a response to 'what did you say four/seven times?' and the deaccented condition as a response to 'did you read or did you say *canto-quanto* four/seven times?' (see [12] and [39] for further details on the association between focus and accent in Italian).

$$
H0 \begin{Bmatrix} \textit{detto QUANTO…CANTO} \\ \textit{DETTO quanto…canto} \end{Bmatrix} \times \begin{Bmatrix} \textit{quattro} \\ \textit{sette} \end{Bmatrix} \textit{volte} \quad \begin{array}{l} \text{'I said\_\_\_four times'} \\ \text{'I SAID\_\_seven times'} \end{array}
$$

Prosodic context   Following word

The stimuli (9 continuum steps × 2 prosodic contexts (accented/deaccented) × 2 following words (*quattro/sette*) × 10 repetitions = 360) were presented to 24 Italian listeners in a two-alternative forced-choice perception test that was conducted online. Participants also heard additional stimuli consisting of the target words *canto-quanto* in isolation; we do not discuss the isolated word data here. Participants were asked to wear headphones and could listen to the stimuli as many times as they wished. Their task was to listen to each phrase stimulus, decide whether the target word sounded more like *canto* or *quanto* and click on the corresponding button. The listener participants were native Italian speakers aged between 19 and 53 years and, in terms of regional variety, all but two were self-reported Standard Italian and/or Tuscan Italian speakers. All participants were paid for their participation with a voucher sent to their email address.

We fitted a generalized linear model within the R package `lme4` with the listener response (2 levels: *quanto*/*canto*) as the dependent variable, prosody (2 levels: accented/deaccented), word (2 levels: *quattro*/*sette*) and the stimulus number as fixed factors, and also included all two-way interactions between these factors. The listener (24 levels) was included as random factor. The significance of any term was obtained by testing whether the full model and one without the term being tested differed significantly from each other.

2.2 Results

We excluded 3/24 listeners from all further analyses because there was no convergence in their derived psychometric curves (i.e. the decision boundaries for these three listeners lay well beyond the range of the stimulus steps).

Our first hypothesis is that the following word (*quattro* vs. *sette*) should influence listeners' decisions and that there should be more *canto* responses when the target word precedes *quattro*. This is because, following Ohala's model, listeners should attribute lip rounding during the target word to anticipatory coarticulation for the upcoming /kw/. But there should be no such bias towards *canto* in the *sette* context since there is no /w/ in the following word to which coarticulation could be attributed. It is clear from Fig. 1—which shows psychometric curves fitted to all 21 listeners separately in the four contexts—that there is no support for this hypothesis. If there had been more *canto* responses preceding *quattro*, then the black curves in Fig. 1 should be to the left of the grey ones. In fact, there appear to be no differences in responses before the two words in the deaccented context while in the accented context, there is even a trend towards more *quanto* responses preceding *quattro*. Thus there is no evidence from this experiment to support the idea that a following /w/ masks the perception of an initial /w/.

The second hypothesis was that there should be an even greater tendency for more *canto* responses before *quattro* in the deaccented condition. Obviously, there is no support for this hypothesis either, given the completely overlapping psychometric curves preceding these words in the deaccented context. On the other hand, the same figure shows that there are very many more *canto* responses in the deaccented than in the accented context. Some suggestions for why this might be so are discussed below.

The observations in Fig. 1 were to a large extent supported by the statistical analysis. Firstly, dropping word as well as both the interaction of word with stimulus and of word with prosody made no significant difference to the statistical model. This result shows that the difference between *quattro*/*sette* had no significant influence on the responses: thus the trend in Fig. 1 by which there were more *quanto* responses pre-

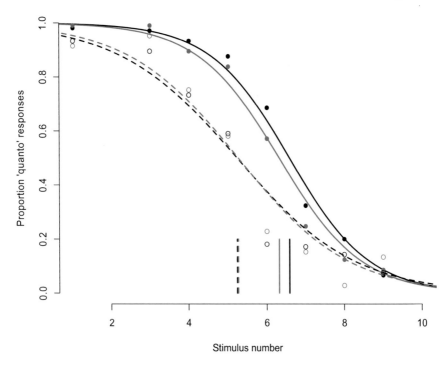

Fig. 1 Fitted psychometric curves showing listener responses to a *quanto-canto* continuum embedded in four contexts: Preceding __*quattro* (*black*) versus __*sette* (*grey*) and in accented (*solid*) versus deaccented (*dashed*) position. The *circles* are the averages of responses across all listeners for any stimulus number. Increasing stimulus numbers are from low to high F2

ceding *quattro* in the accented condition was not supported by the statistical analysis. Consistently with Fig. 1, there was a significant interaction between stimulus number and prosody ($\chi_1^2 = 31.2$, $p < 0.001$): this result shows what is evident in Fig. 1 that there are more *canto* responses along the stimulus continuum in the deaccented versus accented conditions.

2.3 Discussion

We simulated long-range coarticulatory lip-rounding both by lowering F2 in—*anto* and through the presence of /w/ in an immediately following word. The presence of a lip-rounded consonant in the following word (*quattro*) or not (*sette*) made no difference to listeners' responses. We were therefore not able to recreate in the laboratory the sound change by which an initial /w/ in Latin *quinque* dissimilates as a result of a following /w/. This negative finding does not necessarily mean that Ohala's [50] idea about dissimilation through over-compensating for coarticulation is wrong; it may just be that these following word differences are insufficient for recreating in an experiment the conditions by which historic dissimilation could have occurred. But there is, however, an interpretation that could be consistent with Ohala's model based on our second finding that, regardless of the following word, listeners responded far more with *canto* when the target word was deaccented. Our interpretation of this finding is that in deaccented/hypoarticulated speech, the separation between long-range anticipatory lip-rounding from lip-rounding due to the initial /w/ is obscured. Recall that in our stimuli, -*anto* was in all cases synthesised with a very low F2. This makes all the stimuli sound as if they were produced by a speaker with a long-term rounded lip setting. It is this speaker-attribute of lip-rounding that listeners fail to distinguish perceptually from the /w/ of *quanto* (causing them to hear *canto*). Thus the lip-rounding in our data camouflages perceptually the initial /w/. This is entirely consistent with Ohala [48] who also interprets dissimilation as perceptual camouflage.

According to Ohala, the listener error that is the source of dissimilation comes about because of a following /w/. We were not able to demonstrate that with our results. But our results are consistent with the idea that long-range lip-rounding can interfere with the perception of an initial /w/. The further new angle suggested by the present results is that this interference comes about not in all speaking styles, but specifically in hypoarticulated/deaccented speech.

The present study and those in Harrington et al. [28, 29] have suggested a language-internal motivation for sound change which arises because hypoarticulation (simulated here by deaccenting) can be detrimental to parsing coarticulation perceptually. In the next experiment, we consider the extent to which differences at the group level—between adults and children in parsing coarticulation—may additionally contribute to some of the conditions that can cause sound change to occur.

3 The Perception of Coarticulation by Adults and by Children

The coarticulation to be investigated in this experiment was the fronting of the mid-high lax rounded vowel /ʊ/ in a symmetrical /t_t/ context and the acoustic lowering of the mid-high front rounded vowel /ʏ/ in a symmetrical /p_p/ context. The materials were in all cases taken from standard German in which /ʊ, ʏ/ are contrastive (e.g. *musste/müsste*, /mʊste, mʏste/, 'had to'/'should have').

The coarticulatory fronting of /ʊ/ has been extensively documented and comes about because the tongue dorsum for /ʊ/ is shifted forward under the influence of the alveolar constriction [36, 52]. This type of phonetic /ʊ/-fronting causes a raising of its second formant frequency. The coarticulatory F2-lowering in /ʏ/ comes about because the constricted lip gesture for /p/ overlaps with the high front rounded vowel /ʏ/. From another related point of view, the F2-frequency of /ʏ/ is lowered under the influence of the low F2-locus frequency for /p/.

The influence of coarticulation in perception was tested using a well-established technique of embedding an acoustically identical /ʊ-ʏ/ continuum in /t_t/ and /p_p/ contexts and then deriving through a forced-choice listening experiment the cross-over boundary at which responses are equivocal, i.e. at 50 % [21, 46]. The main point to observe here is that the direction in which /ʊ, ʏ/ differ acoustically (from low to high F2) is the same as that of the coarticulatory influence of the /p_p, t_t/ contexts (also from low F2 for /p_p/ to high F2 for /t_t/). Therefore, if listeners adjust their responses in relation to these coarticulatory effects, then they should be more likely to hear /ʊ/ in a /t_t/ than a /p_p/ context. This is further illustrated in Fig. 2 which shows schematically the relationship between production and perception for these vowel × context combinations. The figure shows how the distributions are shifted to the right (towards higher F2) in the production of both vowels in the /t_t/ context than in a /p_p/ context for the reasons stated above: /t/ causes F2 to be raised, and /p/ F2 to be lowered (with the raising effect due to the alveolar possibly being greater than the lowering effect due to the labial). Consequently, if perceptions are adjusted exactly for these effects of coarticulation—that is, if there is 'parity' between the production and perception of coarticulation—then the cross-over boundary in perception from /ʊ/ to /ʏ/ (shown in the lower half of the same figure) should be higher in a /t_t/ than in a /p_p/ context. The issue to be tested is whether this difference in the cross-over boundary (the length of the line marked 'normalise' in Fig. 2) was less for children, which would indicate that they normalise less for coarticulation. This follows from

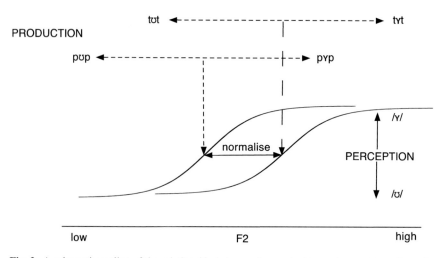

Fig. 2 A schematic outline of the relationship between the production and perception of coarticulation. The *upper part* of the display illustrates the hypothetical distributions in F2 of the four target words showing higher F2 values in a /t_t/ context. The *lower panel* shows the distribution of the corresponding perceptual responses under the assumption that the production and perception of coarticulation are exactly aligned. The degree to which listeners normalise for coarticulation in this model is proportional to the length of the horizontal line marked 'normalise' which extends between the two sigmoids cross-over boundaries at which the probability of perceiving /ʊ/ or /ʏ/ are both 50 %

one of Ohala's [50] predictions that children as less experienced listeners of the language might normalise less for coarticulation than adults.

3.1 Method

There were three parts to the method. Firstly (Sect. 3.1.1), a training phase for the children which also involved the creation of a child-production database. Secondly (Sect. 3.1.2), the creation of perception stimuli to which we obtained forced-choice categorical responses from the adults and imitations (following training) from the children. Thirdly (Sect. 3.1.3), the conversion of the child imitations to categorical responses. These three stages are described more fully below.

3.1.1 Child-Production Database

13 children participated in a training period in which they first learned to associate the target non-words with four puppet names TUTT, TÜTT, PUPP, PÜPP corresponding phonemically to /tʊt, tʏt pʊp, pʏp/ respectively (Fig. 3). Once these had been learned, they produced each of the four puppet names five times. The productions were obtained from the children from a randomised sequence of the puppets' pictures (those in Fig. 3) that were presented on a computer screen one at a time. This child-production database consisted of 4 words × 5 repetitions × 13 children = 260 tokens.

3.1.2 Creation of Synthetic Stimuli

A male speaker of Standard German with slight South German regional characteristics produced utterances containing /pʊp, tʏt/ in the carrier phrase 'Maria hat ___ gesagt' (literally: 'Maria has ___ said') with nuclear accent on the target word. An

Fig. 3 The four puppet pictures used in the picture-naming task by the children for the production (from *left* to *right*) of /tʊt, tʏt, pʏp, pʊp/

11-step F2 continuum was created between original productions of /pʊp/ (F2 = 803 Hz) and /tʏt/ (F2 = 1436 Hz) by using LPC-resynthesis in the static morphing method of Akustyk [56]. The durations of the vowels were normalized using PSOLA. This same 11-step vowel continuum differing in F2 was spliced into labial /p_p/ and alveolar /t_t/ contexts in the same utterance 'Maria hat ___ gesagt' with nuclear accent on the target word (see also [36] for further details).

The stimuli were randomised and presented to a group of 20 L1-German speaking adults (students at the IPS, most of them in their twenties) and to a group of 13 L1-German speaking children (age range from 4 years and 11 months to 6 years and 3 months) resulting in 2 continua (p_p, t_t) × 11 stimuli × 10 repetitions × 20 adult listeners + 2 continua (p_p, t_t) × 11 stimuli × 3 repetitions × 13 children listeners = 5258 presentations. The adults carried out a forced choice identification task and identified each stimulus as one of TUTT, TÜTT, PUPP, PÜPP. Since such a task was considered to be too difficult for the children, they instead imitated each stimulus that they heard following both the training period and the creation of the child-production database as described in Sect. 3.1.1: that is, the children were very familiar with the four characters shown in Fig. 4 before they participated in this perception and imitation experiment.

3.1.3   Obtaining Categorical Responses from Children

Both the child-production (Sect. 3.1.1) and child-imitation (Sect. 3.1.2) corpora were segmented and labelled phonetically and prosodically by two transcribers. The

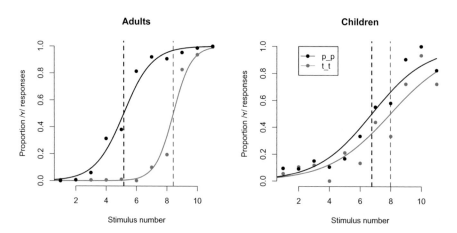

Fig. 4 Psychometric curves showing the proportion of /ʏ/ responses as a function of stimulus number calculated across all subjects for adults (*left*) and children (*right*) in /p_p/ (black) and /t_t/ (*grey*) contexts. The points are proportions at each stimulus number averaged across all subjects (19 adults, 8 children). The *vertical lines* show the 50 % cross-over boundaries at which the proportion of /ʊ, ʏ/ responses are both equal to 0.5. The increasing stimulus numbers extend from low to high F2

acoustic vowel boundaries were marked at the onset and offset of periodicity of each vowel. The formant frequencies were calculated with a 12.5 ms Blackman window and a frame shift of 5 ms. The formant data was checked manually and hand-corrected if necessary. All mis-imitations (e.g. /p/ for /t/ substitutions) and target words that were not phrase-medial were excluded from further analyses (a total of 354 out of 858 tokens).

In order to be able to compare the adult and child data, the child imitations had to be converted into /ʏ/ or /ʊ/ categorical responses. For this purpose, training was carried out on the child-production database and testing involved classifying each imitation as one of these two vowel categories. We included only children who had produced a minimum of 5 /ʊ/ and 5 /ʏ/ vowels, since otherwise it was difficult to achieve statistical stability in constructing the training models. Since three children had produced less than this number of tokens, training and testing were carried out on the data from the remaining 10 children and only the data from those 10 children were further analysed below.

The training and testing were accomplished separately for each child. Training was a Gaussian classification [13, 26, 58] based on the first two formant frequencies at the acoustic temporal midpoint of the vowel. Testing was a maximum likelihood classification based on whichever Bayesian distance to the two vowel categories was smallest.

4 Results

We first removed responses from those combinations of listeners and consonantal-contexts in which there was no convergence in the psychometric curves, i.e. if the resulting decision boundary for any given listener on either continuum fell outside the range of the stimuli (i.e. outside the range of the x-axis shown in Fig. 4). This happened if e.g. a listener responded to a continuum almost entirely with either /ʊ/ or /ʏ/. This required removing all ($n = 6$) response data from 2 children and 1 adult (leaving data from 8 children and 19 adults for further analysis). In addition, responses to the p_p continuum from a further 4 adults and one child ($n = 5$) and to the t_t continuum from additionally 1 child ($n = 1$) had to be removed for the same reason. Thus of the 60 possible original decision boundaries ((20 adults + 10 children) × 2 consonantal contexts), 48 decision boundaries and their associated perceptual responses remained and were analysed below, after removing these data.

As schematically outlined in Fig. 2 above, the greater the distance between the decision boundaries of the psychometric curves, the more listeners normalised for context, i.e. the more they perceived the same acoustic stimulus to be different in the two contexts. The results of the group psychometric curves in Fig. 4 clearly show a greater contextual normalisation for adults than for children. The same figure also shows that the psychometric curves are a good deal flatter for the children which means that they perceived the /ʊ-ʏ/ phonological contrast less distinctively than did adults.

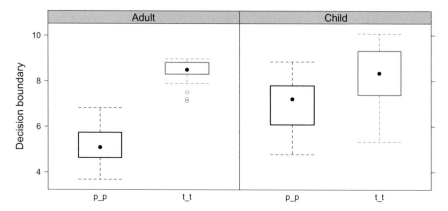

Fig. 5 50% decision boundaries (corresponding to the *vertical dashed lines* in this figure) for 19 adults (*left*) and 8 children (*right*) in the p_p and t_t contexts. There is one data point per listener in each of the four distributions

The individual decision boundaries for the 19 adult and 8 child listeners in Fig. 5 below show, consistently with the group plots in Fig. 4, a much greater separation between the labial and alveolar contexts for adults than for children. Since the results of a mixed model with the decision boundary as the dependent variable, with consonantal context (2 levels: labial, alveolar) and group (2 levels: adult, child) and with the listener as a random factor showed a significant interaction between the fixed factors ($\chi^2_1 = 9.3$, $p < 0.01$), we applied post-hoc Tukey tests to the same data. These showed consistently with Fig. 5 a significant difference between the decision boundaries for adults ($z = 9.1$, $p < 0.001$) but not for children. They also showed a significant difference in the decision boundaries between adults and children in the labial ($z = 3.7$, $p < 0.001$) but not in the alveolar context. Thus as Fig. 5 shows, the decision boundaries in the labial context are much closer to those in the alveolar context for children than for adults.

5 General Discussion

The study has been concerned with the influence of segmental context on the perception of phonological contrasts and the way in which normalisation (compensation) for context can be affected by variation in speaking style (the first experiment) or by differences between speaker groups (the second). The experiments were conceived within the theory being developed in this chapter that parity between how coarticulation is produced and perceived represents a stable association between phonological categories and speech signals, whereas sound change can come about when there is a misalignment between the modalities in processing coarticulation.

The first experiment tested whether such instability was more likely in hypoarticulated signals. The sound change concerned a case of dissimilation by which historical /w/ has been deleted when there is another /w/ that occurs later in the same word. Hypoarticulation was simulated by a post-focal, deaccented synthesis of the target word. The results showed fewer initial /w/ perceptions in the hypoarticulated condition. The fewer perceptions of /w/ in the hypoarticulated condition were not due to the presence or absence of a following /w/ (in the next word), but were instead explained as a result of confusion with the lip-rounding that was simulated synthetically by lowering the second formant frequency throughout the target word. Our interpretation of these results is that in hypoarticulated speech, listeners less effectively parsed the signal into those properties that were due to the initial /w/ and those that came about because of the simulated long-range lip-rounding: that is, in a hypoarticulated speaking style phonetic long-range lip-rounding is more likely to mask or camouflage perceptually the initial /w/ than in hyperarticulated speech.

The second experiment built upon earlier findings by Nittrouer and Studdert-Kennedy [47] to test whether the perceptual adjustments for consonantal context are weaker in children than in adults. Our results were consistent with this hypothesis: adults' decisions were swayed to a greater extent by consonantal context than those of children in categorising German lax high rounded vowels.

We now consider the extent to which our interpretation of these results is consistent with Ohala's [50, 51] theory concerning the conditions under which sound change is likely to occur. In our model, dissimilation resulting from a deletion of the initial /w/ comes about because listeners cannot so easily distinguish the initial /w/ from the effects of long-range lip-rounding in a hypoarticulated speaking style. Thus, in contrast to Ohala, we do not invoke an over-normalisation for coarticulation to explain either these data or the processes leading to dissimilation in general. Further, our model differs from Ohala's because we propose that long-range lip-rounding itself rather than the presence of a second /w/ is sufficient to trigger the perceptual deletion of the first /w/. On the other hand, our interpretation that dissimilation comes about as a result of long-range lip-rounding perceptually masking or hiding a segment that shares the same phonetic properties is very reminiscent of Ohala's [48] view that there is commonality between the mechanisms leading to dissimilation and those in the visual domain causing an object to be camouflaged (as when—to use his analogy—a white rabbit is hidden against a background of snow). The difference here is that our model incorporates speaking style: the perceptual masking that can lead to dissimilation is more likely to happen in response to hypoarticulated speech.

The data from the second experiment are relevant to Ohala's [50] interpretation that many more sound changes arise because of an *insufficient* perceptual normalisation for coarticulation. With regard to the data in the second experiment, Ohala's [50] idea is that the back vowel of /tut/ can change category for the listener into a front vowel /tyt/ when the listener no longer normalises for coarticulation: that is, the prior stage in Ohala's model to diachronic category change is that /u/ in /tut/ is decontexualised such that a listener ceases to apply a perceptual shift to compensate for the phonetic raising effects due to the flanking alveolars. With regard to the model schematised in Fig. 2, Ohala's prediction of perceptual under-compensation for coarticulation is that the right /tʊt-tʏt/ boundary should shift to the left so that the section marked 'normalise' is recategorised as /ʏ/. But this is not what our data show. Recall that there was no difference between children and adults in the location of the /tʊt-tʏt/ boundary: children's diminished normalisation for context instead came about because of a rightwards shift of the /pʊp-pʏp/ boundary towards that of /tʊt-tʏt/.

So do these data from this second experiment have anything to say about sound change? We think that they do. This is because they form a consistent pattern with our other various different types of analyses of /ʊ, u/-fronting based on (i) longitudinal studies in the same individual [25], (ii) an apparent-time study comparing younger and older speakers and listeners [27] in Standard Southern British, and more recently (iii) normalisation for /ʊ/-fronting in deaccented speech in German [28]. Whenever our perception analyses showed evidence for an under-compensation for coarticulation, then, just as for the children in the present experiment, the decision boundary of the /ʊ/ or /u/-variants in the back or non-fronting context shifted to the front—and not as predicted by Ohala's [50] model the other way round. Our findings from perception in these studies are also generally consistent with production data (e.g. for the longitudinal study in Harrington [25]) by which there was a shift of non-fronting (e.g. 'move', /muv/) towards fronted /u/-variants (e.g. 'mute', /mjut/).

Our explanation for these findings across these studies is that diachronic /ʊ, u/ fronting is the outcome of a synchronically gradual shift induced by /ʊ, u/ undershoot which pushes non-fronting ('move') towards fronted ('mute') variants. Acoustically, the shift is brought about by an F2-raising of non-fronting /ʊ, u/-variants due to their centralisation under hypoarticulation. That is, acoustically the F2-space of vowels in words like 'move' becomes more extensive when there is target undershoot/hypoarticulation: crucially, this extension due to hypoarticulation/undershoot is *asymmetric* towards higher F2 values. The perceptual response to this greater asymmetric variation in production is firstly the flatter psychometric response curve observed for the labial context in the present study for adults compared with alveolars (compare the steepness of the two sigmoids in the left panel of Fig. 4). Secondly, if inexperienced listeners are predominantly exposed in speech communication to hypoarticulated back vowels, then it is not just the sigmoid slope of the perceptual contrast that will decrease: the decision boundary will also shift up the F2 scale towards the fronted variant. This we would suggest is the reason why children's perceptual responses in the labial context are shifted much further to the right and nearer to those of the alveolar context than for adults. Based on these data then, diachronic /ʊ, u/-fronting is not brought about as Ohala [50] has argued because a listener gives up compensating for coarticulation (which should cause the alveolar decision boundary to shift towards the labial one). It is instead the outcome of phonologising a back vowel variant ('move') that extends synchronically due to hypoarticulation towards a variant ('mute') that is already front as a result of coarticulation.

The core idea here is then that hypoarticulated variants shift towards those that are substantially affected by coarticulation. To what extent does this idea generalise to other types of sound change? For Ohala [50], sound changes such as diachronic /u/-fronting, umlaut resulting from VCV coarticulation, and the development of vowel nasalisation are all brought about if a listener turned speaker normalises insufficiently for coarticulation. In our model, all of these sound changes are linked not by insufficient compensation for coarticulation, but by the weakening effects induced by hypoarticulation. In the absence of data of our own, we speculate that hypoarticulation can just as easily target the source of coarticulation as the coarticulatory effect, causing its weakening (e.g. modern Standard German *Füße* with a final /ə/ derived from Old High German /fotiz/) or its deletion (as in French 'main', /mɛ̃/ from Latin 'manus'). This weakening/deletion of the source would lead to its decoupling from (and eventual phonologisation of) the coarticulatory effect. Thus these three sound changes share in common that they are all derived not from a perceptual under-compensation for coarticulation but instead from the perceptual consequences of hypoarticulated speech signals: the main difference across these sound changes is that hypoarticulation in /u/-fronting targets the sound that is ultimately changed, whereas in VCV coarticulation and the phonologisation of vowel nasalisation, it targets the source that gives rise to the sound change.

Hypoarticulation is also the synchronic factor that links the two experimental findings of this chapter: in the first, a hypoarticulated speaking style causes a perceptual camouflage between long-range anticipatory coarticulation and an initial consonant with similar phonetic properties, potentially leading diachronically to its dissimilation. In the second, hypoarticulation causes synchronically a bias that shifts back /ʊ/-variants towards their fronted variants, leading potentially to their merger and diachronic recategorisation as front vowels. This merger was evident in the children's responses in Experiment 2. The link between both experiments is the idea that hypoarticulation results in listener uncertainty. This is substantiated by the evidence showing flatter sigmoids for deaccented versus accented responses for adults (Experiment 1: Fig. 1) and for children versus adults (Experiment 2: Fig. 4).

In our proposed model, sound change arises synchronically out of the interaction of the separate forces of hypoarticulation and coarticulation that act upon the transmission of speech. From this point of view, our model integrates the insights expressed respectively in Lindblom et al. [44] and Ohala [50, 51] that both speaking style variation and the perception of coarticulation together set the conditions for sound change to occur. Our model is consistent with past [8, 9, 53, 55] as well as more recent [31, 41, 65] findings that, because lexically more frequent words tend to be hypoarticulated relative to less frequent ones [18, 42, 62], then sound change may often be lexically gradual and dependent on lexical statistics.

Finally, our model in which hypoarticulation plays a central role in setting the conditions for sound change to occur is also consistent with the evidence that sound change is very often reductive. But we emphasise that our model does not predict that sound change *must* be reductive. The data from the second experiment represent just such an example in which diachronic vowel recategorisation emerges out of hypoarticulation pushing (in both perception and production) back towards coarticulatory fronted high vowel variants without any vowel reduction.

Acknowledgments This research was supported by European Research Council grant number 295573 'Sound change and the acquisition of speech' (2012–2017).

References

1. Abrego-Collier, C.: Liquid dissimilation as listener hypocorrection. In: Proceedings of the 37th Annual Meeting of the Berkeley Linguistics Society, pp. 3–17 (2013)
2. Alderete, J., Frisch, S.: Dissimilation in grammar and the lexicon. In: de Lacy, P. (ed.) The Cambridge Handbook of Phonology. Cambridge University Press, Cambridge (2006)
3. Beckman, M., Li, F., Kong, E., Edwards, J.: Aligning the timelines of phonological acquisition and change. Lab. Phonol. **5**, 151–194 (2014)
4. Beddor, P.: A coarticulatory path to sound change. Language **85**, 785–821 (2009)
5. Beddor, P.: Perception grammars and sound change. In: Solé, M.-J., Recasens, D. (eds.) The Initiation of Sound Change. Perception, Production, and Social Factors, pp. 37–55. John Benjamin, Amsterdam (2012)
6. Blevins, J.: Evolutionary Phonology: The Emergence of Sound Patterns. Cambridge University Press, Cambridge (2004)
7. Browman, C., Goldstein, L.: Articulatory phonology: an overview. Phonetica **49**, 155–180 (1992)
8. Bybee, J.: Word frequency and context of use in the lexical diffusion of phonetically conditioned sound change. Lang. Var. Change **14**, 261–290 (2002)
9. Chen, M., Wang, W.: Sound change: actuation and implementation. Language **51**, 255–281 (1975)
10. Cutler, A., Foss, R.: On the role of sentence stress in sentence processing. Lang. Speech **20**, 1–10 (1977)
11. de Jong, K.: The supraglottal articulation of prominence in English: linguistic stress as localized hyperarticulation. J. Acoust. Soc. Am. **97**, 491–504 (1995)
12. D'Imperio, M.: Italian intonation: an overview and some questions. Probus **14**(1), 37–69 (2002)
13. Duda, R., Hart, P., Stork, D.: Pattern Classification, 2nd edn. Wiley, New York (2001)
14. Foulkes, P., Vihman, M.: Language acquisition and phonological change. In: Honeybone, P., Salmons, J.C. (eds.) The Handbook of Historical Phonology. OUP, Oxford (in press)
15. Fowler, C.: Segmentation of coarticulated speech in perception. Percept. Psychophys. **36**, 359–368 (1984)
16. Fowler, C.: Speech as a supramodal or amodal phenomenon. In Calvert, G., Spence, C., Stein, B. (eds.) The Handbook of Multisensory Processes, pp. 189–201. MIT Press, Cambridge (2004)
17. Fowler, C.: Parsing coarticulated speech in perception: effects of coarticulation resistance. J. Phonetics **33**, 199–213 (2005)
18. Fowler, C., Housum, J.: Talkers' signaling of 'new' and 'old' words in speech and listeners' perception and use of the distinction. J. Memory Lang. **26**, 489–504 (1987)
19. Fowler, C., Smith, M.: Speech perception as "vector analysis": an approach to the problems of segmentation and invariance. In: Perkell, J., Klatt, D. (eds.) Invariance and Variability in Speech Processes, pp. 123–139. Erlbaum, Hillsdale (1986)
20. Fowler, C., Thompson, J.M.: Listeners' perception of "compensatory shortening". Attention Percep. Psychophys. **72**, 481–491 (2010)
21. Fujisaki, H., Kunisaki, O.: Analysis, recognition and perception of voiceless fricative consonants in Japanese. Annu. Bull. Res. Inst. Logop. Phoniatr. **10**, 145–156 (1976)
22. Grammont, M.: Observations sur le langage des enfants. In: Mélanges Linguistiques Offerts aI M. Antoine Meillet, pp. 115–131. Klincksieck, Paris (1902)
23. Grosvald, M.: Interspeaker variation in the extent and perception of long-distance vowel-to-vowel coarticulation. J. Phonetics **37**, 173–188 (2009)
24. Grosvald, M., Corina, D.: The production and perception of sub-phonemic vowel contrasts and the role of the listener in sound change. In: Solé, M.-J., Recasens, D. (eds.) The Initiation of Sound Change. Perception, Production, and Social Factors, pp. 77–100. John Benjamins, Amsterdam (2012)

25. Harrington, J.: Evidence for a relationship between synchronic variability and diachronic change in the Queen's annual Christmas broadcasts. In: Cole, J., Hualde, J. (eds.) Laboratory Phonology, vol. 9, pp. 125–143. Mouton, Berlin (2007)
26. Harrington, J.: The Phonetic Analysis of Speech Corpora. Wiley-Blackwell (2010)
27. Harrington, J., Kleber, F., Reubold, U.: Compensation for coarticulation, /u/-fronting, and sound change in Standard Southern British: an acoustic and perceptual study. J. Acoust. Soc. Am. **123**, 2825–2835 (2008)
28. Harrington, J., Kleber, F., Reubold, U.: The effect of prosodic weakening on the production and perception of trans-consonantal vowel coarticulation in German. J. Acoust. Soc. Am. **134**, 551–561 (2013)
29. Harrington, J., Kleber, F., Reubold, U., Siddins, J.: The implications for sound change of prosodic weakening: evidence from polysyllabic shortening. J. Lab. Phonol. **6**, 81–117 (2015)
30. Hay, J., Jannedy, S., Mendoza-Denton, N.: Oprah and /ay/: lexical frequency, referee design and style. In: Proceedings of the 14th International Congress of Phonetic Sciences, San Francisco (1999)
31. Hay, J., Foulkes, P.: The evolution of medial /t/ over real and remembered time. Language (in press)
32. Heid, S., Hawkins, S.:. An acoustical study of long domain /r/ and /l/ coarticulation. In: Proceedings of the 5th Seminar on Speech Production: Models and Data, pp. 77–80. Kloster Seeon, Bavaria, Germany, Munich (2000)
33. Kataoka, R.: Phonetic and cognitive bases of sound change. Ph.D. dissertation, University of California, Berkeley (2011)
34. Kawasaki, H.: Phonetic explanation for phonological universals: the case of distinctive vowel nasalization. In: Ohala, J., Jaeger, J. (eds.) Experimental Phonology, pp. 81–103. Academic Press, Orlando (1986)
35. Kiparsky, Paul: Universals constrain change; Change results in typological generalizations. In: Good, Jeff (ed.) Linguistic Universals and Language Change, pp. 23–53. Oxford University Press, Oxford (2008)
36. Kleber, F., Harrington, J., Reubold, U.: The relationship between the perception and production of coarticulation during a sound change in progress. Lang. Speech **55**, 383–405 (2012)
37. Kleber, F., Peters, S.: Children's imitation of coarticulatory patterns in different prosodic contexts. In: Proceedings of 14th Conference on Laboratory Phonology, Tokyo, Japan (2014)
38. Labov, W.: Sociolinguistic Patterns. University of Pennsylvania Press, Philadelphia (1972)
39. Ladd, D.: Intonational Phonology, 2nd edn. Cambridge University Press, Cambridge (2008)
40. Lightfoot, D.: The Development of Language: Acquisition, Change, and Evolution. Blackwell, Malden (1999)
41. Lin, S., Beddor, P., Coetzee, A.: Gestural reduction, lexical frequency, and sound change: a study of post-vocalic /l/. Lab. Phonol. **5**, 9–36 (2014)
42. Lindblom, B.: Explaining phonetic variation: a sketch of the H & H theory. In: Hardcastle, W., Marchal, A. (eds.) Speech Production and Speech Modeling, pp. 403–439. Kluwer, Dordrecht (1990)
43. Lindblom, B.: Systemic constraints and adaptive change in the formation of sound structure. In: Hurford, J., Studdert-Kennedy, M., Knight, C. (eds.) Approaches to the Evolution of Language: Social and Cognitive Bases, pp. 242–264. Cambridge University Press, Cambridge (1998)
44. Lindblom, B., Guion, S., Hura, S., Moon, S.-J., Willerman, R.: Is sound change adaptive? Rivista di Linguistica **7**, 5–36 (1995)
45. Lindblom B., MacNeilage, P.: Coarticulation: A universal phonetic phenomenon with roots in deep time. FONETIK 2011, 41–44 in TMH—QPSR, vol. 51, KTH Stockholm (2011)
46. Mann, V., Repp, B.: Influence of vocalic context on the perception of [ʃ]-[s] distinction: I. Temporal factors. Percep. Psychophys. **28**, 213–228 (1980)
47. Nittrouer, S., Studdert-Kennedy, M.: The role of coarticulatory effects in the perception of fricatives by children and adults. J. Speech Hear. Res. **30**, 319–329 (1987)
48. Ohala, J.: The listener as a source of sound change. In: Masek, C.S., Hendrick, R.A., Miller, M.F. (eds.) Papers from the Parasession on Language and Behavior, pp. 178–203. Chicago Linguistic Society, Chicago (1981)

49. Ohala, J.: The phonetics of sound change. In: Jones, C. (ed.) Historical Linguistics: Problems and Perspectives, pp. 237–278. Longman, London (1993)
50. Ohala: The listener as a source of sound change: an update. In: Solé, M.-J., Recasens, D. (eds.) The Initiation of Sound Change. Perception, Production, and Social factors, pp. 21–36. John Benjamins, Amsterdam (2012)
51. Ohala, J., Feder, D.: Listeners' identification of speech sounds is influenced by adjacent "restored" phonemes. Phonetica **51**, 111–118 (1994)
52. Öhman, S.E.: Coarticulation in VCV utterances: spectrographic measurements. J. Acoust. Soc. Am. **39**, 151–168 (1966)
53. Phillips, B.: Word Frequency and Lexical Diffusion. Palgrave Macmillan, Basingstoke (2006)
54. Pierrehumbert, J., Hirschberg, J.: The meaning of intonational contours in the interpretation of discourse. In: Cohen, P., Morgan, J., Pollack, M. (eds.) Intentions in Communication, pp. 271–311. MIT Press, Cambridge (1990)
55. Pierrehumbert, J.: Exemplar dynamics: word frequency, lenition, and contrast. In: Bybee, J., Hopper, P. (eds.) Frequency Effects and the Emergence of Lexical Structure, pp. 137–157. John Benjamins, Amsterdam (2001)
56. Plichta, B.: AKUSTYK for Praat [Computer program] (2010). Accessed 12 June 2015
57. Solé, M.: The perception of voice-initiating gestures. Lab. Phonol. **5**, 37–68 (2014)
58. Srivastava, S., Gupta, M., Frigyik, B.: Bayesian quadratic discriminant analysis. J. Mach. Learn. Res. **8**, 1, 277–305 (2007)
59. Studdert-Kennedy, M.: Introduction: the emergence of phonology. In: Hurford, J., Studdert-Kennedy, M., Knight, C. (eds.) Approaches to the Evolution of Language, pp. 169–176. Cambridge University Press, Cambridge
60. Vihman, M.: Sound change and child language. In: Traugott, E., La Brum, R., Shepherd, S. (eds.) Papers from the 4th International Conference on Historical Linguistics, pp. 303–320. John Benjamins, Amsterdam (1980)
61. West, P.: Perception of distributed coarticulatory properties of English /l/ and /r/. J. Phonetics **27**, 405–425 (2000)
62. Wright, R.: Factors of lexical competition in vowel articulation. In: Local, J., Ogden, R., Temple, R. (eds.) Papers in Laboratory Phonology VI, pp. 75–87. Cambridge University Press, Cambridge (2003)
63. Yu, A.: Individual differences in socio-cognitive processing and the actuation of sound change. In: Yu, A. (ed.) Origins of Sound Change: Approaches to Phonologization, pp. 201–227. Oxford University Press, Oxford (2013)
64. Yu, A., Abrego-Collier, C., Sonderegger, M.: Phonetic imitation from an individual-difference perspective: subjective attitude, personality, and 'autistic' traits. PLoS One **8**(9), e74746 (2013)
65. Zellou, G., Tamminga, M.: Nasal coarticulation changes over time in Philadelphia English. J. Phonetics **47**, 18–35 (2014)

Nonlinear Timing and Language Processing in Norm and Pathology

Elzbieta Szelag and Anna Dacewicz

Abstract Many experimental data have indicated that nonlinear temporal processing plays a crucial role in human cognition, including our language communication. We summarize briefly the neuropsychological findings on typical temporal frame underlying language processing. Concentrating on millisecond (the high-frequency processing system) and multisecond (the low-frequency system) timing mechanisms, we provide evidence on temporal dynamics of both speech reception and expression with the special concern on temporal integration and temporal resolution of the signal. The examples of nonlinear temporal processing are discussed in cross-linguistic scenario. Finally, we provide some examples on time distortion in language-disordered population and discuss future applications of training in temporal processing in amelioration of language functions.

Keywords Language · Nonlinear temporal processing · Temporal windows · Cross-linguistic comparisons · Language disorders

1 Introduction

Language is the human learned communication system consisting of arbitrary signs representing the external and internal word, structured according to the grammatical rules. It permits to communicate our perceptions, thoughts and memories. Speech is defined as motor acts—by performing articulatory movements a speaker encodes information and addresses it via acoustic signal to the listener's ear. Thus, the speaker encodes the message which is, next, decoded by the listener. Both the encoding and decoding processes are of a great complexity.

E. Szelag (✉) · A. Dacewicz
Laboratory of Neuropsychology, Nencki Institute of Experimental Biology,
Warsaw, Poland
e-mail: e.szelag@nencki.gov.pl; eszelag@swps.edu.pl

E. Szelag
University of Social Sciences and Humanities, Warsaw, Poland

© Springer International Publishing Switzerland 2016
A. Esposito et al. (eds.), *Recent Advances in Nonlinear Speech Processing*,
Smart Innovation, Systems and Technologies 48,
DOI 10.1007/978-3-319-28109-4_4

Human language has not a monolithic entity and may be separated into several functional sub-systems that control reception (comprehension), expression (production), verbal memory, writing and reading. Each of these functional units is subserved by different mechanisms and processes that are located in different parts of the brain [1]. The specific damage to these structures may have different consequences for our verbal communication. Looking at the dorsal and ventral tracts for language in fiber tracking studies summarized by Friederici and Gierhan [2], it seems important to note that these sub-systems do not work separately and could be described by complex relationships between the components. Moreover, an inherent part of all these sub-systems is a temporal dynamics. Look around—in every language words are composed of sub-sets of sounds (phonemes) which constitute the distinct spectro-temporal units consisting of specific formants (phonological level). The strings of phonemes form syllables, syllables are integrated into words which have meanings (semantics) and they are, next, arranged into meaningful sentences that follow specific rules (syntax). Each of the afore-going levels is characterized by a specific distinct temporal dynamics which may be reflected in a defined temporal analysis, corresponding to some tens of milliseconds (single phonemes), hundreds of milliseconds (syllables), or seconds (sentences or phrases). Thus, the speech signal is characterized by the modulation of sound properties over a wide range of timescales (Fig. 1).

Understanding the mechanisms by which the brain organizes this complex temporal behaviour is a central issue in modern neuroscience. We present here some evidence that, in addition to the central neural control of language, a further level of temporal organization is provided by the nonlinear oscillatory dynamics that are intrinsic to the efferent and afferent pathways. Accordingly, we can produce sequences of oscillatory states that are both spectrally and temporally complex.

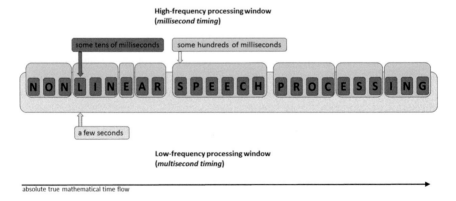

Fig. 1 The relationship between objective (indicated by an *arrow*) and subjective non-linear (indicated by specific time windows) time flow in processing of the speech signal. The millisecond timing corresponds to simple phonemes (*dark grey*) and syllables (*light grey*), whereas the multisecond timing to phrase duration in the fluent speech (*dotted gray*)

Simple variations in such neural signals can result in atypical temporal process-ing leading to deficient speech perception and/or expression. It seems, therefore, that language is inherently an example of a multileveled temporal activity, thus, the use of rhythm, tempo and pausing is crucial for our verbal communication. Such temporal dynamics involves the "higher" level cognitive function in which a global supra-system chunking is implemented.

This paper addresses the nonlinear timing in human verbal communication. We summarize briefly the results of neuropsychological findings on temporal dynamic in language processing in norm and pathology. We begin by describing the typical tem-poral dynamic embedded into language processing. Next, we report some evidence on cross-linguistic universalia observed in different languages, like tonal (Chinese) or phonemic non-tonal (Polish or German), followed by the examples of time dis-tortion in various language-disordered populations. We review predominantly the results of our studies that are illustrated with some literature examples. Finally, we provide some evidence on benefits of training in such nonlinear temporal processing in amelioration of language function.

2 Nonlinear Temporal Processing in a Cross-Linguistic Scenario

The close relationship between temporal information processing and language capa-bilities is well established. Timing in cognitive systems, including language com-munication, is usually considered as nonlinear temporal constraints that comprise the complex relationships between components of an integrated system. It may be characterized as particular steps of typical linguistic processing in both speaking and writing.

Considering the temporal processing in the language domain, we should empha-size one important issue. The concept of general nature of an objective time passing, as well as time continuity was implemented by Isaak Newton and basic physicists. They assumed absolute, true, mathematical time which flows equably without any relation to the external or internal stimulation. In contrast, in cognitive systems, we refer to the subjective flow of time which is characterized by distinct tempo-ral phenomena, reflecting the temporal discontinuity. At least two processing sys-tems (known also as "temporal windows") can be distinguished employing discrete time sampling. Accordingly, one can distinguish a high-frequency processing system related to millisecond timing, as well as a low-frequency processing system charac-terized by a few seconds time domain. We provide below some rationale that these two systems are fundamental for language communication.

The problem of time wrapping of speech is approached using the observation of temporal dynamics in our fluent speech, as well as numerous experimental par-adigms applied to assess time perception abilities in particular individuals. For the high-frequency processing system the most important techniques are temporal gap detection (known also as the fusion threshold paradigm), duration comparison, as well as temporal ordering based on assessment of temporal-order-threshold. This paradigm assesses the smallest temporal gap separating two stimuli presented in rapid sequence that is necessary to report correctly their temporal order, thus, the relation *before–after*. The temporal order threshold reflects the sequencing abilities and may be measured using the following paradigms: (1) monaural stimulus presentation of two clicks or two tone bursts presented to the right and then to the left ear (the identification of relation *left-right* or *right-left*), (2) binaural stimulus presentation comprising two tones differing in frequency (the identification of *low-high* or *high-low* order), and (3) white noise or tone stimuli differing in duration (*short-long* or *long-short*). On the other hand, for the multisecond processing system the most important techniques are time judgement comprising reproduction, production, estimation, or comparison of temporal intervals, moreover, spontaneous reversal time rate of ambiguous figures, and subjective accentuation of metronome beats.

The millisecond timing refers to the duration of single phonemes in the fluent speech. The entire set of formant transitions which build single phonemes, especially stop-consonants like P, B, T, D, K, G has usually ca. 40 ms duration in any language. Another example for the nonlinear temporal constraints can be the Voice-Onset-Time (VOT), introduced by Lisker and Abramson [3]. This paradigm is defined as the time interval between the release of a stop occlusion and the onset of vocal cord vibration: "…the time interval between the burst that marks release and the onset of periodicity that reflects laryngeal vibration…" [3]. Despite cross-linguistic differences in the identification of such phenomena (reflecting positive or negative VOT values), as well as differentiated order of a burst and an onset of vibration in voicing contrast perception across languages, the critical VOT value has usually a duration around some tens of milliseconds. For example in Slavic languages (like Polish), voicing starts well before the release of the plosive (app. 30 ms or more). According to the normative data collected in our Laboratory from 67 healthy Polish speakers, synthesized pseudo-words /Tomek - domek/ (in English: /Tom - the house/) with VOT from −100[1] to −70 ms were usually perceived by as voiced, whereas those with VOT from +5 to +90 ms the unvoiced ones [4]. The categorization pattern in perceptual labelling of voiced and voiceless initial consonants is an example on the existence of nonlinear temporality in speech perception/production.

In case of multisecond processing level we can refer to temporal binding or temporal integration. Such low-frequency processing appears to group successive events into approx. 2–3 s time units. The convincing results indicating such nonlinear processing system comes not only from experimental data on time reproduction, spontaneous rate of ambiguous figures, sensory-motor synchronization, subjective accentuation of metronome beats [5–7] but also from every day observations of the dynamic flow of our fluent speech. In many languages, like Polish, German, English, Russian, Italian or Chinese, verbal utterances of ca. 2–3 s duration are interspersed by short pauses. They are necessary for a speaker to prepare mentally to the next verbal

---

[1]In Slavic languages the negative VOT values are typical for the voicing contrast, whereas, those positive ones for the unvoicing contrast.

unit, as well as for a listener to process the received verbal information. Such temporal chunking is also observed in many languages in poetry, as lines in classic verse often have a duration limited to a few seconds [8]. Accordingly, both the expression and reception of spontaneous flow of speech indicates the temporal chunking limited to a few second. In addition to the high-frequency system (see above), the multisecond timing provides another nonlinear temporal constraints and temporal frame for our language communication.

The other question is whether the mother tongue or the specific language environment could influence millisecond timing mechanisms verified with temporal ordering abilities [9]. To answer this question we performed cross-linguistic comparisons with normal healthy young volunteers of different language background, i.e., Chinese (tonal language users) and Polish (non-tonal, phonetic language users). Using three different measurement procedures focussed on millisecond timing, i.e., monaurally presented clicks ('*left-right*' discrimination) and binaurally presented tones ('*high-low*' discrimination) of either near frequencies (600 and 1200 Hz) or distant frequencies (400 and 3000 Hz) we found that Chinese and Polish subjects have similar temporal order thresholds in 'click task', but significantly reversed threshold patterns in 'tone tasks'. While Chinese subjects have the lower order thresholds for two near frequency tones, Polish subjects demonstrate exactly the opposite—the lower order thresholds for two far frequency tones. Such double dissociation indicates not only a common temporal mechanism for auditory information processing across different language groups evidenced with the 'click task', but also an impact of native language experience on temporal order perception verified with the 'tone tasks'. The long-term exposure to each speech environment creates its own temporal window of optimal processing which is adaptive to its own speech system, but non-adaptive to the other speech system.

3   Temporal Integration Versus Temporal Resolution

The question could be asked why these temporal mechanisms are active when we perceive or produce verbal utterances?

It may be suggested, on one hand, that our focusing on time allows to reduce the complexity imprinting in language communication. Indeed, phonemes are built from formant transitions which are integrated into a global entity within ca. 40 ms. The detection and processing a huge number of incoming features of the arriving signal, e.g., spatio-temporal dimensions of single formants within particular phonemes, could result in overloading of our brains by many details causing less efficient processing. Thus, the temporal mechanisms might allow the dimensionality reduction, fostering optimization of the processing. On the other hand, during speech reception (the afferent pathway) we do not detect usually particular syllables within words, or single words within sentences, but integrate the perceived sounds into meaningful sentences. The similar mechanism may be probably active in the efferent pathway, when the motor program is set in within the premotor cortical circuits, controlling speech production. From the communication point of view, temporal processing may be also related to the rhythmic division of time into the equal portions of verbal utterances, allowing efficient temporal resolution within the signal. It may be responsible for the proper decoding of distinctive acoustic features within the signal.

A number of models for TI have been proposed to answer the question how do listeners integrate temporally distributed verbal information into coherent representations of formants, syllables and words? One hypothesis assumed that sequentially stored items in working memory provide the bottom-up input to unitize the list chunks that group together sequences of items of variable length. The list chunks compete with each other. The winning groupings may create an emergent conscious percept whose properties match the data stored in the reference memory [10].

To sum up, one should distinguish two main issues, i.e., temporal integration and temporal resolution. As stated above, the term temporal integration (TI) refers to summation of features (formants, phonemes, syllables, words) during information processing by the brain. The stimuli may be various types of incoming signals typical for conversation, like speaking, poems, songs, writing and reading. Short succeeding stimuli falling in rapid sequences are fuelled by the need of proper auditory comprehension of the speech signal. It, by its nature, changes rapidly in time. Better understanding of the temporal speech characteristics could help us to improve the communication skill in favourable listening environment what may be especially important in case of language disordered listeners.

On the other hand, the pattern of changes in an acoustic stimulus contains information about the sound source. The message is transmitted by the sender to the listener. Therefore the identification, discrimination, and interpretation of acoustic stimuli depend on the ability of the auditory system to faithfully encode the temporal features of those events. The ability to respond to changes in an acoustic stimulus has been termed temporal resolution (TR). Investigations of TR have focused on the attempt to separate pure temporal from spectro-temporal resolving capabilities. The TR is limited by auditory properties of a listener, related to the hypothetical individual pacemaker. Impaired TR may be conceptualized as a decrease in this smoothing process and, thus, a loss of temporal information. It may be reflected, for example, in poorer decoding of distinctive features or poorer detection of formant transitions characteristic for particular phonemes in the verbal output of speech. The loss of neglected information may result in poorer phonological awareness (deficient phonemic hearing) leading to poorer auditory comprehension of the incoming speech signal. By an analogy, in the efferent pathway during programming and execution of articulatory movement patterns necessary to produce the strings of phonemes deficient TR may result in poorer expression of the rapid transitions within the produced signal.

4 Nonlinear Temporal Processing Deficits in Language Disordered Population

Many experimental data have indicated temporal processing deficits in language disorders in children and adults. The range of temporal information processing is an important factor to consider when studying various subject groups. We provide below some examples on time distortion in such population.

4.1 Acquired Damage to the Brain and Aphasia

Historically, deficient millisecond timing was reported first in aphasic patients by Efron [11] and Swisher and Hirsh [12]. Such deficit was also confirmed in our earlier clinical studies of aphasic patients with post-stroke unilateral focal brain lesions. We observed, however, the interesting relationship between the type of aphasic syndrome and the deficient time range. Deficient millisecond time range was evidenced in Wernicke's aphasics who displayed disordered phonemic hearing and auditory comprehension deficits. In this group deficient timing was evidenced in sequencing abilities measured with the temporal-ordering task. The Wernicke's aphasics needed significantly longer intervals between two successive stimuli to report their order correctly [13]. In contrast, patients with right hemispheric lesions displayed no deficits in temporal order perception, and achieved the order threshold values at the same level as individuals without any brain damage.

On the other hand, Broca's aphasics demonstrated time distortion on multisecond level, verified in the experiment with subjective accentuation of metronome beats [6, 7]. Disordered multisecond system in these patients could be related to poor verbal output, telegraphic speech and shortened phrase length. The dissociation in deficient time range observed in Broca's and Wernicke's aphasics corresponded, thus, to the specific type language of disorders, comprising predominantly either sentence level (Broca's aphasia) or phoneme level (Wernicke's aphasia).

Moreover, in our recent studies [14] we found in aphasic patients significant correlations between the severity of deficient millisecond timing and receptive language deficits which were verified with the Token Test, the Phoneme Discrimination Test, and the Voice-Onset-Time (VOT) Test. These correlation data provide the strong clinical support for the thesis that timing constitutes the core process incorporated in language resources and indicate a clear coexistence of the 'timing - auditory comprehension' relationships.

4.2 Developmental Language Disorders

Deficient millisecond timing was also observed in children with isolated language disorders e.g., Specific Language Impairment (SLI) and dyslexia. SLI is manifested in delayed expressive and receptive language development, despite the normal level of nonverbal intelligence. In dyslexia disturbances concerning reading and writing abilities and are not attributable to overall learning difficulties. Interestingly, it is estimated that these two developmental disorders co-occur up to 50% of cases [15]. In early papers Tallal and co-workers reported that SLI children and dyslexics had difficulties during identification and discrimination of tones presented in rapid succession [16, 17]. It has been assumed, therefore, that difficulties in temporal information processing may constitute the core problem in those language disorders.

SLI. Temporal information processing deficits in SLI children have been also revealed in the recent studies conducted in our Laboratory. Obtained data suggest, that compared to healthy children, SLI children aged from 5 to 8 years need twice as much time between two stimuli to order their relation "*before-after*".

Despite of temporal processing deficits, in the existing literature the other impairments were also taken into consideration as a core deficit in SLI children. Linguistic theories of SLI concentrate mostly on delay in various grammar rules acquisition. Nevertheless, they are usually restricted only to languages with specified grammar structure and do not explain the whole phenomenon of SLI. Another possibility was proposed by McArthur and Bishop [18]. They revealed that frequency discrimination thresholds were elevated in ca. one-third children with SLI, while the duration of the stimuli did not hinder the task performance. Thus, the spectral information processing deficit is observed in a part of a population of SLI children. Presence of both spectral and temporal processing deficits in some SLI children, reported also in electrophysiological studies [19, 20] may support a thesis on nonspecific delay in brain maturation.

Dyslexia. In dyslexic, malfunction of high frequency system may lead to deficit on the level of phonological awareness, e.g., impairment in segmenting the written words into its phonemes. The other theory explaining reading difficulties suggests deficits in the magnocellular visual stream of information. It may be associated with the concept of multisensory temporal binding window [21] in which the visual information is integrated with the auditory one. Such integration time window for binding operations in dyslexic children is probably extended, contrary to the healthy individuals. Thus, the difficulties in dyslexia are associated with difficulties in fitting auditory information (phonemes) with visual information (letters) during reading.

Infantile autism. The clinical observations have suggested that severe abnormalities in social behaviour in autism coexist with disturbances of major cognitive functions, like perception, attention, memory, and/or language). Inadequate communication is observed in severe cases as the absence of speech, or in high functioning children as delayed onset or selective communication deficits. As the temporal dynamics is an essential component of all these cognitive functions, in our study we tested whether children with autism show typical temporal processing on multisecond temporal level. Our focus on this processing level was justified by its importance for every-day motor behaviour, conversational skills and social interactions [22].

Using a temporal-reproduction paradigm, we found that autistic children were unable to link their responses to stimulus duration. Independently of presented stimulus duration, they reproduced auditory or visual stimuli with the same response duration of ca. 3 s on average. The important deficits in duration judgment were accompanied by preserved time window for temporal integration in a residual form. As a few second time window is crucial for many aspects of human mental activity (see above), one might conclude that the existence of temporal integration reported in our study may indicate the residual temporal frame for cognitive resources, despite severe cognitive deficits in autism.

5 Applications of Training in Temporal Processing in Amelioration of Language Function

Nowadays, a rapid expansion of interest among scientists reflects practical approaches to treatment of patients suffering from language disorders. Referring to the evidence on time distortion in language–disordered population, our knowledge on importance of non-linear timing for verbal communication resulted in application of the specific training in temporal processing to improve deficient timing. Because of co-existence of timing and language deficits documented above, it may be expected that such training improves temporal dynamics, resulting in a transfer of improvement form the trained time domain to the untrained language domain. Such hypothesis was verified in our pilot studies conducted in patients suffering from post-stroke aphasia [23, 24] as well as in our recent studies in children with SLI [24]. It should be mentioned that the beneficial effects of temporal training were evidenced not only in language-disordered subjects, but also in normal elderly volunteers indicating age-related cognitive deficits [25]. The modern neuropsychological rehabilitation, therefore, should consider also the training in temporal information processing as a basic component of human cognitive function.

Acknowledgments This research was supported by grant *INNOTECH-K1/IN1/30/159041/ NCBR/12.*

References

1. Gierhan, S.M.E.: Connections for the language in the human brain. Brain. Lang. **127**, 205–221 (2013)
2. Friederici, A.D., Gierhan, S.M.E.: The language network. Curr. Opin. Neurobiol. **23**(2), 250–254 (2013)
3. Lisker, L., Abramson, A.S.: A cross-language study on voicing in initial stops. Acoust. Meas. Word. **20**, 384–422 (1964)
4. Szelag, E., Szymaszek, A.: Test do badania rozumienia mowy u dzieci i dorosłych. Nowe spojrzenie na zegar mózgowy. Gdańskie Wydawnictwo Psychologiczne, Sopot (2014)
5. Pőppel, E.: Pre-semantically defined temporal windows for cognitive processing. Philos. Trans. R. Soc. B. **364**(1525), 1887–1896 (2009)

6. Szelag, E.: Temporal integration of the brain as studied with the metronome paradigm. In: Atmanspacher, H., Ruhnau, E. (eds.) Time, Temporality, Now, pp. 121–131. Springer, New York (1997)
7. Szelag, E., Szymaszek, A., Oron, A.: Aphasia as temporal information processing disorder. In: Vatakis, A., Allman, M. (eds.) Time Distortions in Mind—Temporal Processing in Clinical Populations. Brill, Leiden (2015)
8. Turner F., Pöppel, E.: Metered poetry, the brain and time. In: Rentschler I., Herzberger B., Epstein D. (eds.) Beauty and the Brain. Biological Aspects of Aesthetic, pp. 71–90. Birkhäuser, Basel (1988)
9. Bao, Y., Szymaszek, A., Wang, X., Oron, A., Pőppel, E., Szelag, E.: Temporal order perception of auditory stimuli is selectively modified by tonal and non-tonal language environments. Cognition **129**(3), 579–585 (2013)
10. Gossberg, S., Myers, ChW: The resonant dynamics of speech perception: Interword integration and duration-dependent backward effects. Psychol. Rev. **107**(4), 735–767 (2000)
11. Efron, R.: Temporal perception, aphasia and déjà vu. Brain **86**, 403–424 (1963)
12. Swisher, L., Hirsh, I.J.: Brain damage and the ordering of two temporally successive stimuli. Neuropsychologia **10**, 137–152 (1972)
13. Szalag, E., Pőppel, E.: Temporal perception: a key to understand language. Behav. Brain. Sci. **23**, 52 (2000)
14. Oron, A., Szymaszek, A., Szelag, E.: Temporal information processing as a basis for auditory comprehension: clinical evidence from aphasic patients. Int. J. Lang. Commun. Disord. (2015). doi:10.1111/1460-6984.12160
15. McArthur, G.H., Hoqben, J.H., Edwards, V.T., Heath, S.M., Mengler, E.D.: On the 'specifics' of specific reading disabilities and specific language impairment. J. Child. Psychol. Psychiatry. **41**(7), 869–874 (2000)
16. Tallal, P., Piercy, M.: Developmental aphasia: rate of auditory processing and selective impairment of consonant perception. Neuropsychologia **12**, 83–93 (1974)
17. Tallal, P.: Auditory temporal perception, phonics and reading disabilities in children. Brain Lang. **5**, 167–169 (1980)
18. McArthur, G.M., Bishop, D.V.N.: Speech and non-speech processing in people with specific language impairment: a behavioral and electrophysiological study. Brain Lang. **95**, 260–273 (2005)
19. Benasich, A.A., Choudhury, N., Freidman, J.T., Realpe-Bonilla, T., Chojnowska, C., Gou, Z.: The infant as a prelinguistic model for language impairments: predicting from event-related potentials to behavior. Neuropsychologia **44**, 396–411 (2006)
20. Rinker, T., Kohls, G., Richter, C., Maas, V., Schulz, E., Schecker, M.: Abnormal frequency discrimination in children with SLI as indexed by mismatch negativity (MMN). Neurosci. Lett. **413**, 99–140 (2007)
21. Wallace, M.T., Stevenson, R.A.: The construct of the multisensory temporal binding window and its dysregulation in developmental disabilities. Neuropsychologia **64**, 105–123 (2014)
22. Szalag, E., Kowalska, J., Pőppel, E.: Temporal processing deficits in high-functioning children with autism. Brit. J. Psychol. **95**, 269–282 (2004)
23. Szelag, E., Lewandowska, M., Wolak, T., Seniow, J., Poniatowska, R., Pöppel, E., Szymaszek, A.: Training in rapid auditory processing ameliorates auditory comprehension in aphasic patients: a randomized controlled pilot study. J. Neurol. Sci. **338**(1–2), 77–86 (2014)
24. Szelag, E., Dacewicz, A., Szymaszek, A., Wolak, T., Senderski, A., Domitrz, I., Oron, A.: The application of timing in therapy of children and adults with language disorders. Front. Psychol. in press (2015)
25. Szelag, E., Skolimowska, J.: Cognitive function in elderly can be ameliorated by training in temporal information processing. Restor. Neurol. Neuros. **30**(5), 419–343 (2012)

Temporal Information Processing and Language Skills in Children with Specific Language Impairment

Anna Dacewicz, Kamila Nowak and Elzbieta Szelag

Abstract Deficits in temporal information processing (TIP) have been proposed as a one of the crucial mechanisms underlying language disorders. The aim of the present study was to investigate the relationship between TIP and language skills in children with Specific Language Impairment (SLI). In case of SLI, normal patterns of language acquisition are disturbed, while its etiology is not associated with neurological or environmental factors. In the present paper we assessed in twenty seven SLI children, aged from 5 to 8 years, both the efficiency of TIP and language skills. The result revealed significant correlation between efficiency of TIP and the global language skills. These data may provide one more evidence for the debate on SLI etiology, confirming auditory information processing abnormalities in SLI children.

Keywords Specific Language Impairment · Temporal information processing · Language development

1 Introduction

According to the International Classification of Diseases (ICD-10) [1], Specific Language Impairment (SLI) is manifested in disturbances in normal patterns of language acquisition. The development of language reception or/and expression in children suffering from this deficit is delayed, as compared to that observed in typically developing peers. However, the level of other cognitive functions remains within the normal range. The language disorders in SLI are not directly attributable to neurological, emotional or sensory deficits or environmental factors. Nowadays, the etiology

A. Dacewicz (✉) · K. Nowak · E. Szelag
Laboratory of Neuropsychology, Nencki Institute of Experimental
Biology, Warsaw, Poland
e-mail: a.dacewicz@nencki.gov.pl

K. Nowak · E. Szelag
University of Social Sciences and Humanities, Warsaw, Poland
E. Szelag
e-mail: eszelag@swps.edu.pl

© Springer International Publishing Switzerland 2016   45
A. Esposito et al. (eds.), *Recent Advances in Nonlinear Speech Processing*,
Smart Innovation, Systems and Technologies 48,
DOI 10.1007/978-3-319-28109-4_5

of SLI remains still unclear. Nevertheless, couple potential mechanisms which may constitute the core cause of delayed language development have been a topic of many existing literature studies.

One of the viewpoints implicates deficits in auditory information processing, reflecting problems in encoding both verbal and complex non-verbal auditory information. Historically, Tallal and colleagues suggested that children with SLI may be less efficient in discrimination of speech [2] and non-speech [3] sounds that occurred in rapid succession. Furthermore, one of the factors influencing phonological hearing is the duration of the acoustically discriminable features of presented stimuli [4]. Moreover SLI children displayed also deficits in auditory information processing related to poorer ability of frequency discrimination [5]. Other theories implicate some higher-level difficulties associated with procedural memory deficits [6]. The ambiguity of theories on the core deficit in SLI may be related to the great heterogeneity of this disorder.

1.1 Temporal Information Processing (TIP) in Children with SLI

Deficits in rapid auditory TIP in SLI children were revealed in various tasks: during e.g. identification of the order of paired high- and low-frequency tones presented in rapid succession [7], discrimination of two rapidly presented tones differing in pitch [3], or detection of a brief tone followed immediately by a masking noise [8]. On the contrary, several studies fail to evidence deficient TIP in SLI children [9]. According to McArthur and Bishop [10], inconsistency of the existing results may be associated with the following factors: differences in the applied methodology in particular literature studies, presence of problems in general auditory discrimination ability rather than in isolated deficit in rapid auditory processing, various age of participants in particular studies, characteristics of control groups, or influence of attentional deficits on the measurement of TIP in a given task.

Nevertheless, previous studies revealed that TIP deficits coexist with impaired phonological awareness and verbal short term memory in many children with SLI [11]. This phenomenon may suggest common neural mechanisms that control both verbal and non-verbal auditory processing which may be disordered in SLI children. By analogy a lot of evidences indicates that adult patients with brain injures located in the left hemisphere are impaired in both the perception of temporal order (thus, in TIP) and language [12–15]. Pöppel proposed the central TIP mechanism located in the left hemisphere [16]. This hypothesis may have some references to SLI children. For example, Chiron et al. using dichotic listening task revealed that children with dysphasia do not show predominant activation of the left hemisphere in measurement of cerebral blood flow [17].

Moreover, electrophysiological studies confirmed that both rapid auditory processing and verbal information processing are less mature in children with SLI. Such a viewpoint may be supported by the vast majority of studies which revealed event-related brain potentials (ERP) abnormalities in SLI children. ERPs are electrical potentials recorded from the scalp after stimulus presentation. Benasich et al. revealed that during presentation of two tones in rapid succession, infants from families with a history of language learning impairment showed smaller mismatch response amplitude and longer latency of N250 than children from families without any history of such impairment [18] Furthermore, Davids et al. demonstrated that mismatch negativity ERP was not present in SLI children while it was observed in typically developing peers [19].

Considering this evidence, the aim of the present study was to investigate the relationship between TIP and the level of language skills in Polish children suffering from SLI. We focus here not only on the phonological level of language comprehension, but also on higher degree of syntactic-semantic processing.

2 Method

2.1 Subjects

Participants were twenty seven children suffering from SLI (18 male and 9 female) aged from 5 to 8 years of life. All children were Polish native speakers, right-handed (verified with the Oldfield Questionnaire [20]), had normal level of intelligence (IQ of 85 or higher, measured with the Raven's Coloured Progressive Matrices [21]), and normal hearing level (screening audiometry thresholds equal or lover than 20 dB on 500, 1000, 2000, 4000 Hz frequencies). Children had no neurological or psychiatric diagnosis and they were free from any other neurodevelopmental disorders, like autism, attention deficits or socio-emotional disturbances, as determined by the parental report.

The main inclusion criterion was a developmental language delay, defined as reduced performance measured with the Test for Assessment of Global Language Skills (TAGLS) which constitutes the global assessment of language development in children [22]. All SLI children investigated here displayed the overall standard language score or at least two standard language subtests scores below or equal 4th sten.

2.2 Procedures

Temporal Information Processing Assessment. The paradigm of the auditory Temporal-Order-Threshold (TOT) was applied. TOT is a minimum time gap ʒcp arating two successive stimuli, necessary for a subject to report their temporal order correctly. In this paradigm two stimulus presentation modes were used: monaural and binaural one, implemented into the classical adaptive procedure which was verified in our previous studies [23]. These two modes differed in both physical property of presented stimuli and method of presentation. In both these modes two brief stimuli were presented in rapid succession with various inter-stimulus-intervals (ISI) which varied adaptively in consecutive trials. The stimuli were presented through the headphones at a comfortable listening level. Before the experiment proper, children underwent the introductory session in which the constant ISI of 300 ms duration was applied.

Monaural Presentation Mode. In monaural mode two, 1 ms sounds (rectangular clicks) were presented monaurally in rapid succession, i.e., the first click was presented to one ear followed by the second click to the other ear. Participants were asked to report the order of these two successively presented clicks by pointing the left or the right handset, respectively (corresponding to the first listened click).

Binaural Presentation Mode. In binaural mode, two 10 ms sinusoidal tones of 400 and 3000 Hz were presented binaurally in rapid succession, i.e., the first one was presented to both ears followed by the second tone, presented also to both ears. Children were asked to report the order of two tones by pointing to the response cards.

For every participant outcomes measures, i.e. the TOT values, obtained in these two modes were averaged and the mean auditory TOT value was applied in the further analysis.

Language Assessment. We used TAGLS which is commonly used tool in Poland for assessment of global language skills in children. The TAGLS measures two following language competencies:

- global linguistic skill understood as a general knowledge of the language systems (semantic, syntactic and phonological)
- communicative skills understood as the ability to use language in everyday oral communication.

TAGLS consists of 7 Subscales which are listed below:

1. Story Recall assesses auditory speech comprehension: examiner read aloud a short story and after that the children answers 10 questions related to this story.
2. Lexicon assesses the lexicon capacity: children name 10 pictures and resolve 10 riddles (beginning from "How we called …?").
3. Correction of Sentences assesses the ability to notice and correct both grammar and semantic mistakes in listened sentences.
4. Inflection assesses grammar skills based on the declension ability which is one of the most important grammar features in the Polish language: in this test, children decline nouns combined with adjectives.
5. Asking Questions assesses the ability to create questions based on short stories.

6. Request and Orders assesses level of pragmatic skills: children imagine 5 events from everyday life situations and express requests and orders related to those events.
7. Storytelling assesses the ability to story reproduction: examiner read aloud short story twice and, then, it is reproduced by the child.

3   Results

In SLI children the mean auditory TOT was 195 (SD = 64) ms and the total score on TAGLS was 58 (SD = 21) points. In this paper we concentrate on the correlation analysis concerning the relationship between the skills achieved in TIP (measured with auditory TOT) and language competency. We performed Pearson correlations between the mean TOT from two modes (monaural and binaural) and the total score on TAGLS.

We observed moderate negative correlation ($r = -0.66$; $p < 0.001$, n = 27) between the total score on TAGLS and TOT (Fig. 1). The higher TOT values reflecting the poorer TIP skills coexisted with poorer total score on TAGLS.

It should be noted, however, that the total score on TAGLS did not correlate with child's age or with the IQ level.

The total score possible to obtain in TAGLS is 125 points.

Fig. 1 Scatter data illustrating correlation between TOT values and the language skills assessed with the TAGLS

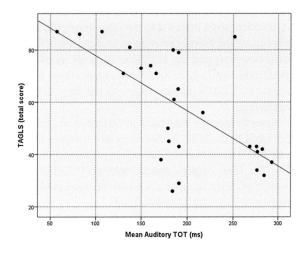

4 Discussion

The present study explored the relationship between TIP and language competency in children with SLI. Our results indicated that delayed language acquisition coexisted with deficits in rapid auditory processing. Children with elevated TOT in both monaural and binaural modes (corresponding to deficient timing) achieved lower outcome measures on TAGLS.

On the basis of previous literature studies, rapid auditory processing is crucial for language comprehension. Acoustic waveform is a complex stream of rapidly changing, acoustic segments varied in frequencies, intensity and temporal characteristic (e.g., formant transitions, Voice-Onset-Time). Hence, for the proper understanding of the speech stream it is necessary to sequencing and splitting properly the acoustic waveform. Since many years, the role of TIP in language processing has been investigated. The vast majority of studies concentrated on time perception in a millisecond time window, corresponding to phoneme level in auditory language comprehension.

Our data are consistent with studies in which poor timing skills were correlated with deficient phonological awareness in children with SLI or dyslexia [11]. However, in our study for the first time, the outcome measures obtained in test which assessed the global language skills turned out significantly correlated with temporal abilities in children. Such result provided a new important evidence on the association between TIP and language. The novel outcome of this study revealed that the global level of language skills is associated with TIP on the millisecond level. It may confirm that TIP is crucial not only for language comprehension, but also for language expression or global communication skills. Therefore, the overall outcomes in TAGLS may reflect a global language deficit. On the other hand, the lower level of language reception may hinder the comprehension of commands in TAGLS in our study. It should be noted however, that the linguistic capacity assessed with TAGLS includes verbal commands, similarly as the other language tests existing in the literature. Thus, the language disorders may overlap with receptive deficits during processing of the test commands.

Nevertheless, the worsened performance on both tasks reported here, thus on TOT and TAGLS, could be mediated by disturbances in attention capacity. In particular, both TOT values and TAGLS scores demanded the elevated level of sustained attention. The level of performance may be vulnerable to the individual sensitivity to distraction. Such interpretation may be supported by literature evidence suggesting comorbidity of ADHD and SLI [24].

The most important finding of the present study is that our data may have some practical applications in predicting benefits of language therapy based on improvement of TIP. TIP measurement may constitute a valuable tool to differentiate the etiology of the delay in language development. It may also imply the most effective solution for language therapy. Accordingly, the conventional speech therapy in children with difficulties in rapid auditory information processing may be less efficient, as it does not address the core problem of an individual, i.e., an impairment of auditory patterns perception in spoken language rather than of motor patterns in speech production. So far, the existing studies reported that such therapies ameliorated language function in children with SLI [25], children with reading difficulties [26] and aphasic patients [14]. Moreover not only language function, but also other cognitive functions may be ameliorated after the temporal training [27].

5 Conclusion

To sum up, the present study provides the evidence that the level of language development is correlated with millisecond timing not only in terms of language comprehension but also on the broader aspects of language skills.

Acknowledgments This research was supported by grant *INNOTECH-K1/IN1/30/159041/ NCBR/12*. The authors want to thank Monika Kastory-Bronowska† and Elżbieta Chruścicka from the Early Childhood Intervention Centre in Warsaw for recruitment of children with SLI to the present study. The authors would like also to thank Dr. Aneta Szymaszek, M.Sc. Ing. Anna Bombińska and our graduate students for their contribution to data collection.

References

1. World Health Organization: The ICD-10 Classification of Mental and Behavioural Disorders: Clinical Descriptions and Diagnostic Guidelines. World Health Organization, Geneva (1992)
2. Tallal, P., Piercy, M.: Developmental aphasia: rate of auditory processing and selective impairment of consonant perception. Neuropsychologia **12**, 83–93 (1974)
3. Tallal, P., Piercy, M.: Deficits of non-verbal auditory perception in children with developmental aphasia. Nature **241**(5390), 468–469 (1973)
4. Tallal, P., Piercy, M.: Developmental aphasia: the perception of brief vowels and extended stop consonants. Neuropsychology **13**, 69–74 (1975)
5. McArthur, G.M., Bishop, D.V.M.: Which people with specific language impairment have auditory processing deficits? Cogn. Neuropsychol. **21**, 79–94 (2004)
6. Ullman, M.T., Pierpont, E.I.: Specific language impairment is not specific to language: the procedural deficit hypothesis. Cortex **41**, 399–433 (2005)
7. Tallal, P., Stark, R., Kallman, C., Meltis, D.: A reexamination of some nonverbal perceptual abilities of language-impaired and normal children as a function of age and sensory modality. J. Speech Hear. Res. **24**, 351–357 (1981)
8. Wright, B.A., Lombardino, L.D., King, W.M., Puranik, C.S., Leonard, C.M., Merzenich, M.M.: Deficits in auditory temporal and spectral resolution in language-impaired children. Nature **387**, 176–178 (1997)
9. Schulte-Körne, G., Deimel, W., Bartling, J., Remschmidt, H.: Role of auditory temporal processing for reading and spelling disability. Percept. Mot. Skill **86**, 1043–1047 (1998)
10. McArthur, G.M., Bishop, D.V.M.: Auditory perceptual processing in people with reading and oral language impairments: current issues and recommendations. Dyslexia **7**, 150–170 (2001)
11. Vandewalle, E., Boets, B., Ghesquière, P., Zink, I.: Auditory processing and speech perception in children with specific language impairment: relations with oral language and literacy skills. Res. Dev. Disabil. **33**, 635–644 (2012)
12. Efron, R.: Temporal perception, aphasia and déjà vu. Brain **86**, 403–424 (1963)

13. von Steinbüchel, N., Wittmann, M., Strasburger, H., Szelag, E.: Auditory temporal-order judgement is impaired in patients with cortical lesions in posterior regions of the left hemisphere. Neurosci. Lett. **264**(1–3), 168–171 (1999)
14. Szelag, E., Lewandowska, M., Wolak, T., Seniow, J., Poniatowska, R., Pöppel, E., Szymaszek, A.: Training in rapid auditory processing ameliorates auditory comprehension in aphasic patients: a randomized controlled pilot study. J. Neurol. Sci. **338**(1–2), 77–86 (2014)
15. Oron, A., Szymaszek, A., Szelag, E.: Temporal information processing as a basis for auditory comprehension: Clinical evidence from aphasic patients. Int. J. Lang. Commun. Disord. **50**(5), 604–615 (2015)
16. Pöppel, E.: A hierarchical model of temporal perception. Trends Cogn. Sci. **1**(2), 56–61 (1997)
17. Chiron, C., Pinton, F., Masure, M.C., Duvelleroy-Hommet, C., Leon, F., Billard, C.: Hemispheric specialization using SPECT and stimulation tasks in children with dysphasia and dystrophia. Dev. Med. Child Neurol. **41**, 512–520 (1999)
18. Benasich, A.A., Choudhury, N., Freidman, J.T., Realpe-Bonilla, T., Chojnowska, C., Gou, Z.: The infant as a prelinguistic model for language impairments: predicting from event-related potentials to behavior. Neuropsychologia **44**, 396–411 (2006)
19. Davids, N., Segers, E., van den Brink, D., Mitter, H., van Balkom, H., Hagoort, P., Verhoeven, L.: The nature of auditory discrimination problems in children specific language impairment: an MMN study. Neuropsychologia **49**, 19–28 (2011)
20. Oldfield, R.C.: The assessment and analysis of handedness: the Edinburgh Inventory. Neuropsychologia **9**, 97–113 (1971)
21. Szustrowa, T., Jaworowska, A.: Manual for Raven's Progressive Matrices: Coloured Progressive Matrices. Polskie Towarzystwo Psychologiczne, Warszawa (2003)
22. Tarkowski, Z.: Test Sprawności Językowej. Orator, Lublin (2001)
23. Szymaszek, A., Sereda, M., Pöppel, E., Szelag, E.: Individual differences in the perception of temporal order: the effect of age and cognition. Cogn. Neuropsychol. **26**(2), 135–147 (2009)
24. Mueller, K.L., Tomblin, J.B.: Examining the comorbidity of language disorders and ADHD. Top. Lang. Disord. **32**(3), 228–246 (2012)
25. Merzenich, M.M., Jenkins, W.M., Johnston, P., Schreiner, C., Miller, S.L., Tallal, P.: Temporal processing deficits of language-learning impaired children ameliorated by training. Science **271**, 81–84 (1996)
26. Hook, P.E., Macaruso, P., Jones, S.: Efficacy of Fast ForWord Training on facilitating acquisition of reading skills by children with reading difficulties–a longitudinal study. Ann. Dyslexia **51**, 75–96 (2001)
27. Szelag, E., Skolimowska, J.: Cognitive function in elderly can be ameliorated by training in temporal information processing. Restor. Neurol. Neurosci. **30**(5), 419–434 (2012)

Kinematic Modelling of Dipthong Articulation

Cristina Carmona-Duarte, Jesús B. Alonso, Moises Diaz,
Miguel A. Ferrer, Pedro Gómez-Vilda and Réjean Plamondon

Abstract The Sigma-Lognormal model has successfully been applied to handwriting modeling but never to acoustic-phonetic articulation. The hypothesis of this paper is that vocal tract dynamics, which includes jaw and tongue can be approached by the Kinematic Theory. In speech analysis, the movement of the tongue and jaw has been linked to the variation on first and second formants. In this paper, we explore the Kinematic hypothesis, based on diphthong pronunciation, which invoke the most extreme tongue gestures in the vowel triangle, estimation of their formants, and transformation of these to space for evaluating the speed profile. The estimated speed profile is modelled by the sigma lognormal model of the Kinematic Theory. An average reconstruction error of 20 dB has been obtained in the experiments carried out with 20 different volunteers. This result validates the work hypothesis, opening a new research line in speech processing.

Keywords Sigma-Lognormal · Formant · Kinematic · Diphthong · Articulation

C. Carmona-Duarte (✉) · J.B. Alonso · M. Diaz · M.A. Ferrer
Instituto Universitario para el Desarrollo Tecnológico y la Innovación en Comunicaciones,
Universidad de Las Palmas de Gran Canaria, Las Palmas de Gran Canaria, Spain
e-mail: ccarmona@idetic.eu

J.B. Alonso
e-mail: jalonso@idetic.eu

M. Diaz
e-mail: mdiaz@idetic.eu

M.A. Ferrer
e-mail: mferrer@idetic.eu

P. Gómez-Vilda
Facultad de Informática, Universidad Politécnica de Madrid,
Campus de Montegancedo, s/n, 28660 Boadilla del Monte, Madrid, Spain
e-mail: pedro@fi.upm.es

R. Plamondon
Laboratoire Scribens, Déartement de Génie Électrique,
École Polytechnique de Montréal, Montreal, Canada
e-mail: réjean.plamondon@polymtl.ca

© Springer International Publishing Switzerland 2016
A. Esposito et al. (eds.), *Recent Advances in Nonlinear Speech Processing*,
Smart Innovation, Systems and Technologies 48,
DOI 10.1007/978-3-319-28109-4_6

1   Introduction

It is well known in speech analysis that a specific vowel phonation depends on the velo-pharyngeal switch, the tongue neuromotor system, the mandibular system and the laryngeal system. In the case of vowels, a space representation can be obtained from the two first formants (F1 and F2) [1]. For instance, in [2] the jaw-tongue dynamics is considered as the basis to explain the space formant distribution in running speech. Besides, [2] shows a simple model to estimate jaw-tongue dynamics from the two first formants.

The Kinematic Theory of rapid human movements [3] describes the way in which neuromuscular systems are involved in the production of muscular movements. This theory has been applied successfully to handwriting, analyzing the neuromuscular system involved in the production of rapid movements [4], the variations of handwriting with time [5, 6], the prevention of brain strokes [7], the specification of a diagnostic system for neuro-muscular disorder [9], etc.

In the case of handwriting, the arm and trunk muscles act to generate the handwriting. Similarly, the tongue and jaw muscles are moving to generate different vowels sounds during the speech articulation. So, our hypothesis includes that both speech and handwriting signals could be studied as a human movement. Furthermore, a parallelism can be established between both signals.

In the present paper, a method to estimate the speed profile from the two first formants is introduced, using the Sigma-Lognormal model to foresee the possibility of its applicability to model speech dynamics.

The paper is organized as follows: in Sect. 2 an introduction to the physiology of phonation and the method used to transform formants into distance is presented. In Sect. 3, a brief description of the Sigma-Lognormal model is introduced. In Sect. 4, we present the method followed to estimate the speed signal from the two first formants. The results of the experiments carried out over 20 subjects are presented. Finally, conclusions are commented in Sect. 5.

2   Physiology of Phonation

The articulatory organs and nasal cavity allow focusing the energy of the speech signal at certain frequencies (formants), due to oropharyngeal tract resonators. Estimating the resonance or formant structure of voiced speech is possible from a digital inverse filter formulation. There are numerous techniques to perform the inverse filtering of a speech signal as the Iterative Adaptive Inverse Filtering algorithm [9], which can provide an adequate estimation of the glottal excitation. However, a linear prediction model based on an autoregressive process (AR) [10] is enough to determine the formants (in non-nasal phonations). In this case, speech signal $s(n)$ can be modelled as follows:

$$s(n) = \sum_{i=1}^{N_{LP}} a_{LP}(i)s(n-i) + e(n) \tag{1}$$

where N_{LP} represents the order of the predictor, $\{a_{LP}\}$ are the coefficients of linear prediction (LPC) and $e(n)$ represents the error in the model. The LPC coefficients are calculated by least squared error algorithms and define the transfer function of the vocal tract, V(z), assumed to be given as an all-pole function:

$$V(z) = \frac{G}{1 - \sum_{i=1}^{N_{LP}} a_{LP}(i)z^{-i}} = \frac{G}{\prod_{i=1}^{N_{LP}} (1 - p_{LP}(i)z^{-1})} \tag{2}$$

The poles characterize the formants which are the local maxima of the spectrum, where the first (F1) and second (F2) formants correspond to the two first maximum values in the LPC spectrum. The estimation of the predictor order is based on the sample frequency: for an fs of 22050 samples/s a compromise is to use 15 coefficients.

Also, it is well known that in the vowel phonation the formants F1 and F2 vary for each vocals creating a vowel triangle [1] (Fig. 1).

The acoustic representation spaces are associated to jaw position, articulation place and lip rounding. Assuming a simplification in the acoustic representation, the closed-open gesture and a back-front gesture can be defined. The first is produced by the muscles involved in the jaw movements and the second by the tongue movements. These movements can be correlated with the formants positions in plane F1 versus F2 [2] as:

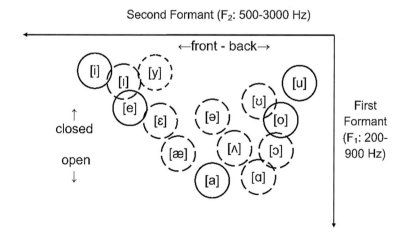

Fig. 1 Vowel representation spaces adapted from [1]. Spanish (*full circle*) and American English (*long-dash circle*)

$$\begin{bmatrix} \Delta x \\ \Delta y \end{bmatrix} = \begin{bmatrix} c_{11} & c_{12} \\ c_{21} & c_{22} \end{bmatrix} \begin{bmatrix} \Delta F1 \\ \Delta F2 \end{bmatrix} \tag{3}$$

where Δx and Δy are the relative displacement from an initial position of the jaw and tongue (x_0, y_0). c_{ij} are the weights of the combination matrix.

3 Overview of the Sigma-Lognormal Model

In this work, we applied a Sigma-Lognormal model [11] in order to parameterize the speed profile of the tongue movement. This model considers the resulting speed of a neuromuscular system action or stroke describing a lognormal function scaled by a command (D) and time-shifted by the time occurrence of the command (t_0) [11]. The complex pattern is produced by summing each resulting lognormal function, given by equation:

$$\overrightarrow{v_n}(t) = \sum \overrightarrow{\bigwedge}(t) = \sum_{i=1}^{M} \vec{v}_j(t) \tag{4}$$

where M represents the number of simple movements involved in the generation of a given pattern, $\bigwedge(t)$ is the Sigma-Lognormal and $\vec{v}_j(t)$ is the velocity profile of the jth stroke.

The speed in the Cartesian space can be calculated as:

$$v_{n\_x}(t) = \sum_{i=1}^{M} |\vec{v}_j(t)| \cos(\emptyset_j) \tag{5}$$

$$v_{n\_y}(t) = \sum_{i=1}^{M} |\vec{v}_j(t)| \sin(\emptyset_j) \tag{6}$$

where $\emptyset_j$ is the direction angle in the jth stroke.

Given v_{n_x} and v_{n_y}, the goodness of their reconstruction from the Sigma-Lognormal domain, is given by the error between the original and its reconstructed signal, which must be as minimum as possible. This criterion can be evaluated using the Signal-to-Noise-Ratio (SNR) between the reconstructed speed profile $(\vec{v}_v(t))$ and the original one $(\vec{v}_v(t))$. In this way, the SNR is defined as:

$$SNR = 20 \log \left(\frac{\int_{t_s}^{t_n} [v_{x\_n}^2(t) + v_{y\_n}^2(t)] dt}{\int_{t_s}^{t_n} [\left(v_{x\_n}^2(t) + v_x^2(t) \right)^2 + (v_{yn}^2(t) - v_y^2(t))^2] dt} \right) \tag{7}$$

The higher is the SNR the better is the reconstruction. Generally speaking, a SNR greater than 20 dB provides excellent signal reconstruction.

4 Methods

4.1 Subjects

Twenty mid age healthy subjects (fifteen males and five females) participated in the experiment. They were recorded during the utterance of three different diphthong phonations pronounced by Spanish speakers (/au/, /iu/, and /ai/).

4.2 Formant Estimation

Formants F1 and F2 for each sample (see Fig. 2) were estimated. Then, Eq. 3 was used to transform the two formants into a space representation. For this purpose, the study in [12] was taken as a reference to estimate a first approximation to the coefficients. In Fig. 3, we can observe the special representation of the Δx and Δy obtained from the F1 and F2 from one subject and the three analyzed diphthongs.

4.3 Speed Profile

The speed profile $\vec{v}(t)$ from the calculated Δx and Δy was estimated with:

$$|\vec{v}(t)| = \sqrt{\Delta x^2 + \Delta y^2} \qquad (8)$$

The resulting speed profile can be seen in Fig. 4.

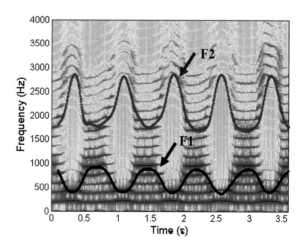

Fig. 2 Formant estimation for diphthong /ai/

Fig. 3 Formant to position
transformation

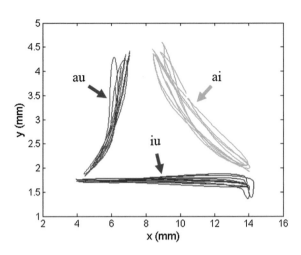

Fig. 4 Speed profile for /ai/
phonation

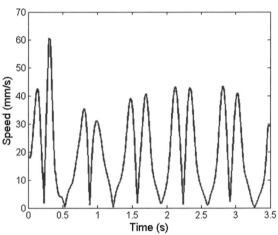

Finally, the quality of the reconstruction from formants to Sigma-Lognormals was evaluated. For such, The Sigma-Lognormal model was used in order to parameterize the resulting speed profile. Then, the SNR (Eq. 7) between the reconstructed signal and the original one was calculated. The results of the experiment are presented in the next section.

5   Experiments and Results

In the experiment, 4 s long utterances of the three diphthongs (/ai/, /iu/ and /au/) were recorded at 22,050 Hz and 16 bits resolution. As it was explained in previous

Fig. 5 Corresponding speed profile (*full line*) and its Sigma-Lognormal decomposition (*dot line*)

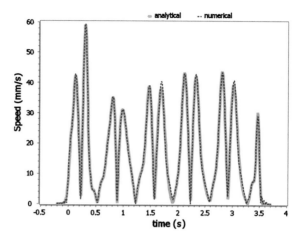

Fig. 6 Signal-to-Noise-Ratio for each diphthong

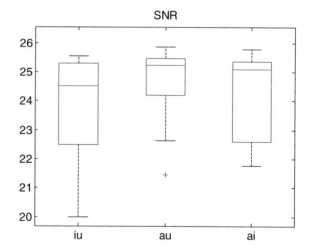

section, formants positions F1-F2 were extracted each 5 ms and the speed profile was estimated.

Once the speed profile is calculated, the reconstructed speed profile is extracted automatically, using the Sigma-Lognormal model explained briefly in Sect. 3 [11]. A typical original speed profile and its Sigma-Lognomal reconstruction one are shown in Fig. 5. It can be observed how close to the original profile the reconstructed speed profile is. In this particular example the SNR is 25 dB.

The SNR was estimated from Eq. 7. The results are shown in Fig. 6. As it is seen from these results, the SNR is greater than 20 dB, which assesses the excellence of fitting the original speed with the Kinematic Theory as in the case of handwriting [6].

6 Conclusions

In this paper, the possibility of modeling formant dynamics as a function of the articulation organ position representation space has been introduced and evaluated. In order to validate the work hypothesis, the original speed profile was estimated from the distance representation of the formants. This speed profile has been reconstructed using the Kinematic Theory and its associated Sigma Lognormal model. The results obtained from the experiments, grant an estimation fit showing a signal-noise ratio of more than 20 dB, which could be considered as an excellent validation figure of this hypothesis.

These first results open new ways to model speech dynamics and the possibility of applying these advances to handwriting, and to the study of neurodegenerative speech production.

Acknowledgments This study was funded by the Spanish government's MCINN TEC2012-38630-C04 research project and the fellowship program of Universidad de Las Palmas de Gran Canaria.

References

1. Peterson, G.E., Barney, H.L.: Control methods used in a study of the vowels. J. Acoust. Soc. Am. **24–2**, 175–184 (1952)
2. Gómez-Vilda, P., Londral, A.R.M., Rodellar-Biarge, V., Ferrández-Vicente, J.M., de Carvalho, M.: Monitoring amyotrophic lateral sclerosis by biomechanical modeling of speech production. Neurocomputing **151**, 130–138 (2014)
3. Plamondon, R.: A kinematic theory of rapid human movements: part I: movement representation and generation. Biol. Cybern. **72**, 295–307 (1995)
4. Plamondon, R., Djioua, M., Mathieu, P.: A time-dependence between upper arm muscles activity during rapid movements: observation of the proportional effects predicted by the kinematic theory. Hum. Mov. Sci. **32**, 1026–1039 (2013)
5. Gomez-Barrero M., Galbally J., Plamondon R., Fierrez J., Ortega Garcia J.: Variations of handwritten signatures with time: a Sigma-Lognormal analysis. In: Proceedings 6th International conference on Biometrics, pp. 3.16.1–3.16.6. Madrid, Spain, 4–7 June 2013
6. Plamondon, R., O'Reilly, C., Rémi, C., Duval, R.C.: The lognormal handwriter: learning, performing and declining. Front. Psychol. Cogn. Sci. (2013). doi:10.3389/fpsyg.2013.00945, Special Issue in Cognitive Science, Writing words: From brain to hand(s), Topic Editor(s): Sonia Kandel, Marieke Longcamp, 1–14
7. Plamondon, R., O'Reilly, C., Ouellet-Plamondon, R.: Strokes against strokes: strokes for strides. Pattern Recognit. **47**, 929–944 (2014)
8. O'Reilly, C., Plamondon, R.: Design of a neuromuscular disorders diagnostic system using human movement analysis. In: 11th International Conference on Information Sciences, Signal Processing and their Applications, Montreal, Canada, 3–5 July 2012
9. Alku, P.: Glottal wave analysis with pitch synchronous iterative adaptive inverse filtering. Speech Commun. **11**(2), 109–118 (1992)
10. Itakura, F.: Line spectrum representation of linear predictor coefficients of speech signals. J. Acoust. Soc. Am. **57**(S1), S35–S35 (1975)
11. O'Reilly, C., Plamondon, R.: Development of a Sigma-Lognormal representation for on-line signatures. Pattern Recognit. **42**, 3324–3327 (2009)
12. Dromey, C., Jang, G.O., Hollis, K.: Assessing correlations between lingual movements and formants. Speech Commun. **55**, 315–328 (2013)

Part III
Identifying Psychological, and Neural Disorders from Speech

Multi-class Versus One-Class Classifier in Spontaneous Speech Analysis Oriented to Alzheimer Disease Diagnosis

K. López-de-Ipiña, Marcos Faundez-Zanuy, Jordi Solé-Casals, Fernando Zelarin and Pilar Calvo

Abstract Most of medical developments require the ability to identify samples that are anomalous with respect to a target group or control group, in the sense they could belong to a new, previously unseen class or are not class data. In this case when there are not enough data to train two-class One-class classification appear like an available solution. On the other hand non-linear approaches could give very useful information. The aim of our project is to contribute to earlier diagnosis of AD and better estimates of its severity by using automatic analysis performed through new biomarkers extracted from speech signal. The methods selected in this case are speech biomarkers oriented to Spontaneous Speech and Emotional Response Analysis. In this approach One-class classifiers and two-class classifiers are analyzed. The use of information about outlier and Fractal Dimension features improves the system performance.

Keywords One-class classifier · Nonlinear Speech Processing · Alzheimer disease diagnosis · Spontaneous Speech · Fractal Dimensions

K. López-de-Ipiña (✉) · P. Calvo
Universidad Del País Vasco/Euskal Herriko Unibertsitatea, Europa Pz 1,
20008 Donostia, Spain
e-mail: karmele.ipina@ehu.eus

P. Calvo
e-mail: pilarmaria.calvo@ehu.eus

M. Faundez-Zanuy
Fundació Tecnocampus, Avda. Ernest Lluch 32, 08302 Mataró, Spain
e-mail: faundez@tecnocampus.cat

J. Solé-Casals
Data and Signal Processing Research Group, University of Vic–Central
University of Catalonia, Sagrada Família 7, 08500 Vic, Spain
e-mail: jordi.sole@uvic.cat

F. Zelarin
GuABIAN, Association for Personal Autonomy,
Avda Zarautz, 6, 4 left, 20018 Donostia, Spain
e-mail: guabiangipuzkoa@gmail.com

© Springer International Publishing Switzerland 2016
A. Esposito et al. (eds.), *Recent Advances in Nonlinear Speech Processing*,
Smart Innovation, Systems and Technologies 48,
DOI 10.1007/978-3-319-28109-4_7

63

1   Introduction

Many applications (most of medical developments) require the ability to identify samples that are anomalous with respect to a target group or control group, in the sense they belong to a new, previously unseen class or are not class data as in not common diseases or environment with very few population. In this case there are not enough data to train two-class models classifier, as in pilot studies, one possible approach to this kind of verification or identification problems is one-class classification, learning a description of the target class concerned based solely on data from this class. One-class classification problem [1] differs from multi-class classifier in one essential aspect. In one-class classification it is assumed that only information of one of the classes, the target class, is available. This means that just example objects of the target class can be used and that no information about the other class of outlier objects is present. The different terms such as fault detection, anomaly detection, novelty detection and outlier detection originate from the different applications to which one class classification can be applied. The boundary between the two classes has to be estimated from data of only control class. The task is to define a boundary around the target class, such that it accepts as much of the target/control objects as possible, while it minimizes the chance of accepting outlier objects. In the literature a large number of different terms have been used for this problem. The term one-class classification originates from [2]. The application is as follows: The first application for one class classification (also called data description as it forms the boundary around the whole available data) is outlier detection, to detect uncharacteristic objects from a dataset. Secondly, data description can be used for a classification problem where one of the classes is sampled very well, while the other class is severely under sampled. The measurements on the under sampled class might be very expensive or difficult to obtain. Finally, the last possible use of the outlier detection is the comparison of two data sets. Assume that a classifier has been trained (in a long and difficult optimization process) on some (possibly expensive) data [3]. As explained above, the second application of outlier detection is in the classification problem where one of the classes is sampled very well but it is very hard and expensive, if not impossible, to obtain the data of the second class. One of the major difficulties inherent in the data (as in many medical diagnostic applications) is this highly skewed class distribution. The problem of imbalanced datasets is particularly crucial in applications where the goal is to maximize recognition of the minority class.

Alzheimer's Disease (AD) is the most common type of dementia among the elderly. It is characterized by progressive and irreversible cognitive deterioration with memory loss and impairments in judgment and language, together with other cognitive deficits and behavioral symptoms. The cognitive deficits and behavioral symptoms are severe enough to limit the ability of an individual to perform everyday professional, social or family activities. As the disease progresses, patients develop severe disability and full dependence. An early and accurate diagnosis of AD helps patients and their families to plan for the future and offers the best opportunity to treat the symptoms of the disease. According to current criteria, the diagnosis is expressed with different degrees of certainty as possible or probable AD when dementia is present and other possible causes have been ruled out. The diagnosis of definite AD requires the demonstration of the typical AD pathological changes at autopsy [4–6]. In addition to the loss of memory, one of the major problems caused by AD is the loss of language skills. We can meet different communication deficits in the area of language, including aphasia (difficulty in speaking and understanding) and anomia (difficulty in recognizing and naming things). The specific communication problems the patient encounters depend on the stage of the disease [6–8].

The main goal of the present work is feature search in Spontaneous Speech Analysis (ASSA) an Emotional Response Analysis (ERA) oriented to pre-clinical evaluation for the definition of test for AD diagnosis. These features will define control group (CR) and AD disease. Non-invasive Intelligent Techniques of diagnosis may become valuable tools for early detection of dementia. Moreover, these techniques are very low-cost and do not require extensive infrastructure or the availability of medical equipment. They are thus capable of yielding information easily, quickly, and inexpensively [9, 10]. This study is focuses on early AD detection and its objective is the identification of AD in the pre-clinical (before first symptoms), Mild Cognitive Impairment (MCI) and prodromic (some very early symptoms but no dementia) stages. The research presented here is a complementary preliminary experiment to define thresholds for a number of biomarkers related to spontaneous speech. Feature search in this work is oriented to pre-clinical evaluation for the definition of test for AD diagnosis. Obtained data will complement the biomarkers of each person [11]. In addition to the loss of memory, one of the major problems caused by AD is the loss of language skills (Fig. 1, [12]). We can meet different communication deficits in the area of language, including aphasia (difficulty in speaking and understanding) and in Emotional Response [13–16].

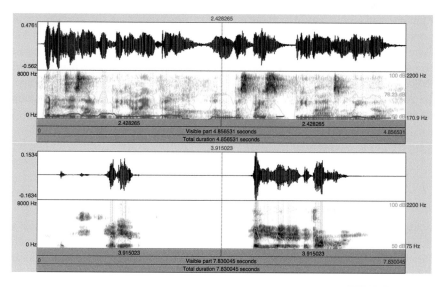

Fig. 1 Signal and spectrogram of a control subject (*top*) and a subject with AD (*bottom*) during Spontaneous Speech (pitch in *blue*, intensity in *yellow*)

2 Materials

This study is focused on early AD detection and its objective is the identification of AD in the pre-clinical (before first symptoms) and prodromic (some very early symptoms but no dementia) stages. The research presented here is a complementary preliminary experiment to define thresholds for a number of biomarkers related to spontaneous speech. Feature search in this work is oriented to pre-clinical evaluation for the definition of test for AD diagnosis. Obtained data will complement the bio-markers of each person. Trying to develop a new methodology applicable to a wide range of individuals of different sex, age, language and cultural and social background, we have built up a multicultural and multilingual (English, French, Spanish, Catalan, Basque, Chinese, Arabian and Portuguese) database with video recordings of 50 healthy and 20 AD patients (with a prior diagnosis of Alzheimer) recorded for 12 h and 8 h, respectively. The age span of the individuals in the database was 20–98 years and there were 20 males and 20 females. This database is called AZTIAHO. A subset of 20 AD patients was selected (68–96 years of age, 12 women, 8 men) with a distribution in the three stages of AD as follows: First Stage [ES = 4], Secondary Stage [IS = 10] and Tertiary stage [AS = 6]. The control group (CR) was made up of 20 individuals (10 male and 10 female, aged 20–98 years) representing a wide range of speech responses. This subset of the database is called AZTIAHORE [10].

3 Methods

In previous work [10, 16] the goal of the experimentation was to examine the potential of the selected features to help in the automatic measurement of the degradation of Spontaneous Speech, Emotional Response and their manifestation in people with AD as compared to the control group. The approach's performance was very satisfactory and promising results for early diagnosis and classification of AD patient groups but medical doctors propose new experimentation oriented to detect mainly early stage. The goal of this new experimentation is to detect changes with regard to CR group and outliers which point to presence of AD's symptoms. One class classification will be used for this propose.

3.1 Feature Extraction

Spoken language is one of the most important elements defining an individual's intellect, social life, and personality; it allows us to communicate with each other, share knowledge, and express our cultural and personal identity. Spoken language is the most spontaneous, natural, intuitive, and efficient method of communication among people. Therefore, the analysis by automated methods of Spontaneous Speech

(SS—free and natural spoken communication), possibly combined with other methodologies, could be a useful non-invasive method for early AD diagnosis. The analysis of Spontaneous Speech fluency is based on three families of features (SSF set), obtained by the Praat software package (Praat) and software that we ourselves developed in MATLAB. For that purpose, an automatic Voice Activity Detector (VAD) has extracted voiced/unvoiced segments as parts of an acoustic signal [12, 17].

3.1.1 Linear Features

These three families of features include:

1. Duration: histogram of voiced and unvoiced segments, the average voiced/unvoiced, and variations.
2. Time domain: short time energy;
3. Frequency domain, quality: spectral centroid.

In this study, we aim to accomplish also the automatic selection of emotional speech by analyzing three families of features in speech:

1. Acoustic features: pitch, standard deviation of pitch, max and min pitch, intensity, standard deviation of intensity, max and min intensity, period mean, period standard deviation, and Root Mean Square amplitude (RMS);
2. Voice quality features: shimmer, local jitter, Noise-to-Harmonics Ratio (NHR), Harmonics-to-Noise Ratio (HNR) and autocorrelation;
3. Duration features: locally voiceless frames, degree of voice breaks.
4. Emotional Temperature [10].

3.1.2 Fractal Dimension

Most of the fractal systems have a characteristic called self-similarity. An object is self-similar if a close-up examination of the object reveals that it is composed of smaller versions of itself. Self-similarity can be quantified as a relative measure of the number of basic building blocks that form a pattern, and this measure is defined as the Fractal Dimension. It should be noted that the Fractal Dimension of natural phenomena is only measurable using statistical approaches. Consequently, there exists no precise reference of the Fractal Dimension value that a given waveform should have. In addition, speech waveforms are not stationary, so most ASR techniques employ short sections of the signal in order to extract features from the waveform. This means that one plausible technique for extracting features from speech waveforms, for the purpose of recognizing different phonemes, is to divide the signal in short chunks and calculate the features for each chunk. This was the approach we adopted. In other words, we calculated the Fractal Dimension of short segments of the waveform and observed the evolution of the obtained values along the whole signal, with the aim of finding in it fractal characteristics that could help in identifying different elements of the spoken message. There are several algorithms for measuring the Fractal Dimension. In this current work we focus on the alternatives which are especially suited for time series analysis and which don't need previous modelling of the system. One of these algorithms is Higuchi [18] named from his author. Higuchi and Castiglioni were chose because it has been reported to be more accurate in previous works with under-resourced conditions [10, 19]. Higuchi [18] proposed an algorithm for measuring the Fractal Dimension of discrete time sequences directly from the time series $x(1), x(2), \ldots, x(n)$.

Without going into detail, the algorithm calculates the length $L_m(k)$ (see Eq. 1) for each value of m and k covering all the series.

$$L_m(k) = \frac{\sum_{i=1}^{\lfloor \frac{N-m}{k} \rfloor} |x(m+ik) - x(m+(i-1)k)|(n-1)}{\lfloor \frac{N-m}{k} \rfloor k} \tag{1}$$

After that, a sum of all the lengths $L_m(k)$ for each k is determined with Eq. 2.

$$L(k) = \sum_{m=1}^{k} L_m(k) \tag{2}$$

And finally, the slope of the curve $\ln(L(k))/\ln(1/k)$ is estimated using least squares linear best fit, and the result is the Higuchi Fractal Dimension (HFD).

The selection of an appropriate window size to be used during the experiments is essential. Broadly speaking, the Fractal Dimension is a tool for attempting to capture the dynamics of the system. With a short window, the estimation is highly local and adapts fast to the changes in the waveform. When the window is longer, some details are lost but the Fractal Dimension better anticipates the characteristics of the signal. Additionally, previous studies that take into account the window size of similar dimension estimations [20–22] suggest that a bigger window could be useful in some cases. Consequently, four window-sizes of 160, 320 and 1280 points will be analyzed.

3.1.3 Features Sets

In the experimentation, three families of feature sets will be used:

1. SSF+EF: Spontaneous Speech features and Emotional Speech features
2. SSF+EF+ET: SSF+EF and Emotional Temperature
3. SSF+EF+ET+VHD: SSF+EF, Emotional Temperature, Higuchi Fractal Dimension and its maximum, minimum, variance, standard deviation

3.1.4　Automatic Classification

The main goal of the present work is feature search in Spontaneous Speech oriented to pre-clinical evaluation for the definition of tests for AD diagnosis. These features will define CR group and the three AD levels. Moreover a secondary goal will be the optimization of computational cost oriented to real time applications. Thus automatic classification will be modeled in this sense. Two different classifiers will be evaluated:

1. Multi-class classifier: Multi Layer Perceptron (MLP) with Neuron Number in Hidden Layer (NNHL) = max(Attribute/Number+Classes/Number) and Training Step (TS) NNHL*10.
2. One-Class classifier: The base classifiers to be used in the experimentation were Bagging and MLP.

WEKA [23] software was used in carrying out the experiments. The results were evaluated using Classification Error Rate (CER) and Accuracy (Acc). For the training and validation steps, we used k-fold cross-validation with k = 10. Cross validation is a robust validation for variable selection [24].

4　Results and Discussion

The task was Automatic Classification, with the classification targets being: healthy speakers without neurological pathologies and speakers diagnosed with AD. The experimentation was carried out with AZTIAHORE. The results have been analysed with regard to the feature set described in Sect. 3. Experimentation has been divided to test One-class classifier only with speech samples from CR group and with information about outliers (patients with AD) and Multi-class classifier (MLP). The results are shown in Figs. 2, 3. In preliminary experiments and based in previous works a window-size of 320 samples have been selected. The results are satisfactory for this study because they obtained very good results for all feature sets (Fig. 2.). Most of them have about 100 % of Accuracy for CR group. However it must be highlighted that despite global result are optimums there is a lack for experiment without outlier. In this case results are very poor for people with AD because there are not enough samples to train properly the MLP. The best results are obtained for SSF+EF+ET+VHFD.

One-class classifier outperforms MLP when outliers group is not used in the training process. The results are satisfactory for this study in the case of MLP classifier and outliers. Bagging paradigm presents lower computational cost but very poor results. The new fractal features improve the system but they can improve Bagging performance. The best results are obtained for integral feature set, which mixes features relative to Spontaneous Speech and Emotional Response (SSF+VHFD+EF+ET). ET appears also as a powerful feature to discriminate AD. The results are satisfactory for this study because they obtained very good results not only for MLP classifier

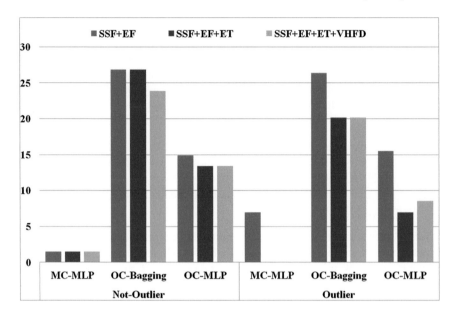

Fig. 2 %CER global results for Multi-class and One-class classifiers

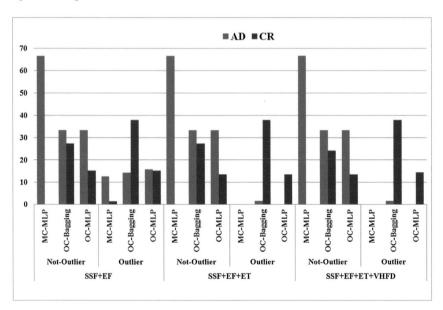

Fig. 3 %CER classes' results for Multi-class and One-class classifiers

but also for Bagging, which presents lower computational cost. One-class classifier with MLP is the option which shows optimum computational cost and better performance in all cases. Moreover this model out-performs MLP when there is a lack of appropriate AD samples.

5 Conclusions

One-class classification has been used for detection of AD symptoms in spontaneous speech with under-resourced condition. In one-class classification it is assumed that only information of one of the classes, the target class, is available. In this work we explore the problem of imbalanced datasets by using information about outlier and, Fractal Dimension and Emotional Temperature features. This option improves the system performance. One-class classifier with MLP outperforms two-class classifier when there are very few AD samples. The approach of this work complements the previous multi-class modelling (two class) and is robust in terms of capturing the dynamics of the speech, and it offers many advantages in terms of be easier to compare the power of the new features against the previous ones. In future works we will introduce new features relatives to speech modelling oriented to standard medical tests for AD diagnosis and to emotion response analysis.

Acknowledgments This work has been supported by FEDER and Ministerio de ciencia e Innovación, TEC2012-38630-C04-03, the University of the Basque Country by EHU-14/58 and the University of Vic–Central University of Catalonia by R0947. The described research was performed in laboratories supported by the SIX project; the registration number CZ.1.05/2.1.00/03.0072, the operational program Research and Development for Innovation.

References

1. Tax, D.M.J.: One-class classification, Ph.D. thesis, Technische Universiteit Delft (2001)
2. Moya, M., Koch, M., Hostetler, L.: One-class classifier networks for target recognition applications. In: Proceedings world congress on neural networks, pp. 797–801. International Neural Network Society, INNS, Portland, OR (1993)
3. Khan, S.S., Madden, M.G.: A Survey of Recent Trends in One Class Classification. Lecture Notes in Computer Science, pp. 188–197. Springer, Berlin (2010)
4. Mc Kahn, G, et al.: Clinical diagnosis of Alzheimer's disease: report of the NINCDS-ADRDA Workgroup on AD, **24**, 939-944 (1984)
5. McKhann, G.M., et al.: The diagnosis of dementia due to Alzheimer's disease: recommendations from the NIAA's association workgroups on diagnostic guidelines for AD. Alzheimers Dement. **7**(3), 263–269 (2011)
6. van de Pole, L.A., et al.: The effects of age and Alzheimer's disease on hippocampal vol-umes, a MRI study. Alzheimer's and Dementia **1**(1, Supplement 1), 51 (2005)
7. Morris, J.C.: The clinical dementia rating (CDR): current version and scoring rules. Neu-rology **43**, 2412b–2414b (1993)
8. American Psychiatric Association: Diagnostic and Statistical Manual of Mental dis-orders, 4th edn. Text Revision, Washington DC (2000)

9. Faundez-Zanuy, M., et al.: Biometric applications related to human beings: there is life beyond security. Cogn. Comput. (2012). doi:10.1007/s12559-012-9169-9
10. Lopez-de-Ipiña, K., Alonso, J.B., Travieso, C.M., Solé-Casals, J., Egiraun, H., Faundez-Zanuy, M., Ezeiza, A., Barroso, N., Ecay, M., Martinez-Lage, P., Martinez-de-Lizardui, U.: On the selection of non-invasive methods based on speech analysis oriented to automatic alzheimer disease diagnosis. Sensors 13(5), 6730–6745 (2013a)
11. Alzheimer's Association. http://www.alz.org/
12. Praat: doing phonetics by computer. www.fon.hum.uva.nl/praat
13. Kwon, O.W., Chan, K., Hao J., Lee, T.W.: Emotion recognition by speech signals. In: Proceedings of 8th European Conference on Speech Communication and Technology (ECSCT'03), pp. 125–128. Geneva, Switzerland, 1–4 September 2003
14. Buiza, C.: Evaluación y tratamiento de los trastornos del lenguaje. Matia Fundazioa, Donostia (2010)
15. Martinez, F., Garcia, J., Perez, E., Carro, J., Anara, J.M.: Patrones de prosodia expresiva en pacientes con enfermedad de alzheimer. Psicothema 24(1), 16–21 (2012)
16. Lopez-de-Ipiña, K., Alonso, J.B., Solé-Casals, J., Barroso, N., Faundez-Zanuy, M., Travieso, C., Ecay-Torres, M., Martinez-Lage, P., Egiraun, H.: On automatic diagnosis of alzheimer's disease based on spontaneous speech analysis and emotional temperature. Cogn. Comput. Springer, Berlin (2013b). doi:10.1007/s12559-013-9229-9
17. Solé-Casals, J., Zaiats, V.: A non-linear VAD for noisy environment. Cogn. Comput. 2(3), 191–198 (2010)
18. Higuchi, T.: Approach to an irregular time series on the basis of the fractal theory. Physica D 31277, 283 (1988)
19. Ezeiza A., López de Ipiña, K., Hernández, C., Barroso, N.: Enhancing the feature extraction process for automatic speech recognition with fractal dimensions, Cogn. Comput. pp. 1–6, Springer-Verlag. (2912), (2013). doi:10.1007/s12559-12-91-65.0
20. Tsonis, A.: Reconstructing dynamics from observables: the issue of the delay parameter revisited. Int. J. Bifurca. Chaos 17, 4229–4243 (2011)
21. Jang, J.S.R.: Audio signal processing and recognition. In: Roger Jang's Homepage (2011). http://www.cs.nthu.edu.tw/~jang, Accesed in (2011)
22. Esteller, R., Vachtsevanos, G., Echauz, J., Litt, B.: A comparison of waveform fractal dimension algorithms. IEEE Trans. Circuits Syst. I Fundam. Theory Appl. 2001 48(2), 177–183 (2012)
23. WEKA. http://www.cs.waikato.ac.nz/ml/weka/
24. Picard, R., Cook, D.: Cross-validation of regression models. J. Am. Stat. Assoc. 79(387), 575–583 (1984). doi:10.2307/2288403.JSTOR2288403

On the Significance of Speech Pauses in Depressive Disorders: Results on Read and Spontaneous Narratives

Anna Esposito, Antonietta M. Esposito, Laurence Likforman-Sulem,
Mauro N. Maldonato and Alessandro Vinciarelli

Abstract This paper investigates whether and how depressive disorders affect speech and in particular timing strategies for speech pauses (empty and filled pauses, as well as, phoneme lengthening). The investigation is made exploiting read and spontaneous narratives. The collected data are from 24 subjects, divided into two groups (depressed and control) asked to read a tale, as well as, spontaneously report on their daily activities. Ten different frequency and duration measures for pauses and clauses are proposed and have been collected using the PRAAT software on the speech recordings produced by the participants. A T-Student test for independent samples was applied on the collected frequency and duration measures in order to ascertain whether significant differences between healthy and depressed speech measures are observed. In the "spontaneous narrative" condition, depressed patients exhibited significant differences in: the average duration of their empty pauses, the average frequency, and the average duration of their clauses. In the read narratives, only the average pause's frequency of the clauses was significantly lower in the depressed subjects with respect to the healthy ones. The results suggest that depressive disorders affect speech quality and speech production through pause and clause

A. Esposito (✉)
Department of Psychology, Seconda Università di Napoli, and IIASS, Naples, Italy
e-mail: iiass.annaesp@tin.it

A.M. Esposito
Istituto Nazionale di Geofisica E Vulcanologia, Sezione di Napoli,
Osservatorio Vesuviano, Naples, Italy
e-mail: antonietta.esposito@ingv.it

L. Likforman-Sulem
Telecom ParisTech, Paris, France
e-mail: laurence.likforman@telecom-paristech.fr

M.N. Maldonato
Department of European and Mediterranean Cultures,
Università della Basilicata, Potenza, Italy
e-mail: m.maldonato@unibas.it

A. Vinciarelli
School of Computing Science, University of Glasgow, Glasgow, UK
e-mail: Alessandro.Vinciarelli@glasgow.ac.uk

© Springer International Publishing Switzerland 2016   73
A. Esposito et al. (eds.), *Recent Advances in Nonlinear Speech Processing*,
Smart Innovation, Systems and Technologies 48,
DOI 10.1007/978-3-319-28109-4_8

durations, as well as, clause quantities. In particular, the significant differences in clause quantities (observed both in the read and spontaneous narratives), suggest a strong general effect of depressive symptoms on cognitive and psychomotor functions. Depressive symptoms produce changes in the planned timing of pauses, even when reading, modifying the timing of pausing strategies.

Keywords Depressive disorders · Speech pauses · Filled and empty pauses · Vowel and consonant lengthening · Clauses · Pausing strategies · Phonation time · Silences

1 Introduction

Speech structure is characterized by pauses, the role and cognitive dimension of which were first underlined by Rochester [33], Butterworth [7], Chafe [10], Clark and Fox Tree [11] and more recently by Esposito et al. [17] and Benus et al. [4], attributing to them the function of signaling complex cognitive planning processes related to the effort to communicate new information content.

Coarsely, speech pauses have been described as a multi-determined phenomenon serving different communicative social functions, among these, synchronizing the verbal and nonverbal communication modes [18, 19], marking the boundaries of narrative units [27–30], predicting a person's status [32], as well as, identifying the caller/receiver's status in phone calls, the caller doing significantly more pauses than the receiver [37].

Pauses in speech are distinguished into three coarse categories as "empty", and "filled" pauses and "consonant/vowel lengthening" (this first categorization was due to Maclay and Osgood [23]). Empty pauses are defined as silent speech intervals, the length of which may vary depending on the task under consideration [15, 16]. Filled pauses are vocalizations such as *uhm, uh, ehm*, etc. Some authors tend to attribute different communicative functions to empty and filled pauses [9]. For example, it is supposed that empty pauses are mostly used to signal phrase boundaries while filled pauses serve to keep the speaker's turn [11]. The lengthening of vowels and/or consonants in word final or central positions is conceptualized as due to pre-articulatory planning processes [24]. Clauses are considered by definition "a sequence of words grouped together on a semantic or functional basis". Clauses are tied to pausing strategies because they are segmented through the three abovementioned different pause categories.

The proposed investigation aims at identifying speech pause duration and pause's frequency measures robust enough to show significant differences between depressed and healthy speech. Such measures will serve for the development of algorithms implemented in automatic diagnostic tools for the early detection and diagnosis of depressive disorders.

2 Methods

2.1 Subjects

The 12 depressed patients are Italian and were recruited with the help of psychiatrists at the Department of Mental Health and the General Hospital in Caserta, the Institute for Mental Health and the General Hospital in Santa Maria Capua Vetere, the Centre for Psychological Listening in Aversa (Italy) and in a private psychiatric office. They were already diagnosed as depressed and some of them were under treatment. In addition they were administered the Beck Depressive Inventory Second Edition (BDI II) proposed by Beck et al. [3] in the Italian language version assessed by Ghisi et al. [20]. Table 1 reports their code (because of anonymity constraints all the subjects received a subject identification code when enrolled in the experiment and were identified by it), gender, age, and BDI II scores. Before being enrolled in the experiment, informed consensus were gathered from all of them. Moreover, the subjects were informed by the experimenter that the aims and the expectations of the proposed research were to find voice and face (also facial expressions were recorded, even though, not for all the involved participants) features that may serve for detecting depressive states. No mention to speech pauses was indeed made.

The speech was first collected for the depressed patients and then the control group was matched to them for age and gender. The BDI II was also administered to the healthy group. Table 1 reports both the experimental and control group demographic variables. Participants were local inhabitants of the Campania Region (Italy), and therefore they matched for geographical/cultural areas and social rule sharing.

2.2 Procedure and Experimental Set-up

Participants were conducted to a quiet room and seated in front of a PC. They were provided with headphones and asked to provide their speech in two different experimental conditions. In the former, they were asked to spontaneously report on their weekly activities. In the latter, they were asked to read from the PC monitor the tale "The North Wind and the Sun" by Esopo. The tale is a standard phonetically balanced short folk tale (about six sentences all together), frequently used in phoniatric practices.

The recordings were made using a clip-on microphone (Audio-Technica ATR3350), with external USB sound card. Speech was sampled at 16 kHz and quantized at 16 bits. For each subject, the recording procedure did not last more than 15 min.

Table 1 Demographic variables describing the experimental (depressed) and control group paired by age

| Experimental group | | | | | | | Control group | | | |
CODE	GENDER	AGE	BDI II Scores	Diagnosis	Personality	TREATMENT	CODE	GENDER	AGE	BDI II Scores
3	M	60	33	Severe	Bipolar	yes	44	M	55	2
17	M	53	20	Moderate		no	6	M	49	5
18	M	29	25	Moderate		yes	42	M	33	2
4	F	63	27	Moderate		no	39	F	60	1
15	F	49	20	Moderate		yes	36	F	39	4
13	F	48	16	Mild		no	32	F	48	5
12	F	46	31	Severe		no	10	F	46	4
24	F	41	16	Mild		no	34	F	40	12
8	F	35	36	Severe		no	35	F	36	0
9	F	33	38	Severe		yes	33	F	36	9
14	F	27	19	Mild	Bipolar	no	40	F	27	8
19	F	27	25	Moderate		no	30	F	27	10

2.3   Measurements

The recordings were analyzed through the PRAAT (http://www.fon.hum.uva.nl/praat/) software, which allows to display at the same time both the spectrogram and waveform of each recording. The pauses (either filled, or empty or consonant/vowel lengthening) were identified manually through their spectral characteristics (from the spectrogram) and then boundaries were identified through the listening of these spectral chunks. The details of the criteria applied to identify the boundaries in the speech waveform are accurately described in [14].

The 10 measurements taken on each participant' (both depressed and healthy subjects) recordings (both read and spontaneous narratives) are described as follow:

1. Empty pause rate1 (EPR1): It was calculated as the ratio between the number of empty pauses (frequency) and the total duration of each spontaneous and read narrative (measured in seconds);
2. Empty pause rate2 (EPR2): It was calculated as the ratio between the total duration of empty pauses and the total duration of each spontaneous and read narrative;
3. Filled Pause Rate1 (FPR1): It was calculated as the ratio between the number of filled pauses (frequency) and the total duration of each spontaneous and read narrative;
4. Filled Pause Rate2 (FPR2): It was calculated as the ratio between the total duration of filled pauses and the total duration of each spontaneous and read narrative;
5. Lengthening Rate1 (LR1): It was calculated as the ratio between the number of consonant/vowel lengthening (frequency) and the total duration of each spontaneous and read narrative (in seconds);
6. Lengthening Rate2 (LR2): It was calculated as the ratio between the total duration of consonant/vowel lengthening and the total duration of each spontaneous and read narrative;
7. Clause Rate1 (CR1): It was calculated as the ratio between the number of clauses (frequency) and the total duration of each spontaneous and read narrative;
8. Clause Rate2 (CR2): It was calculated as the ratio between the total duration of clauses and the total duration of each spontaneous and read narrative;
9. Pause Rate1 (PR1): It was calculated as the ratio between the sum of EPR1, FPR1, and LR1 (that are frequencies) and the total duration of each spontaneous and read narrative;
10. Duration Pause Rate1 (DPR1): It was calculated as the ratio between the sum of empty, filled pauses, and consonant/vowel lengthening durations and the total duration of each spontaneous and read narrative.

For sake of clarity, Table 2 exemplifies the description of the above measurements.

Table 2 Description and acronyms assigned to the measurements under examination

Duration in s/duration in s	Acronym	Frequency/duration in s	Acronym
Empty pause durations/Speech total duration	*EPR2*	Empty pause frequency/Speech total duration	**EPR1**
Filled pause durations/Speech total duration	**FPR2**	Filled pause frequency/Speech total duration	**FPR1**
Lengthening durations/Speech total duration	**LR2**	Lengthening frequency/Speech total duration	**LR1**
Pause total durations/Speech total duration	*DPR1*	Pause total frequency/Speech total duration	**PR1**
Clause durations/Speech total duration	*CR2*	*Clause frequency/Speech total duration*	*CR1*

3   Results

A T-Student test for independent samples was applied on the collected measures in order to ascertain significant differences between healthy and depressed speech.

In the "spontaneous narratives" it was found that:

(a) DPR1, the total duration of speech pauses (empty, filled and vowel lengthening taken all together) is significantly longer ($\alpha < 0.05$) for depressed subjects with respect to healthy ones, under one (T(22) = 2.438257, $\rho = 0.011646$) and two tailed testing hypotheses (T(22) = 2.438257, $\rho = 0.023293$);
(b) EPR2, the total duration of empty pauses is significantly longer ($\alpha < 0.05$) for depressed subjects with respect to healthy ones under one (T(22) = 2.282619, $\rho = 0.016239$) and two tailed testing hypotheses (T(22) = 2.282619, $\rho = 0.032479$);
(c) CR2, the clause duration is significantly shorter ($\alpha < 0.05$), for depressed subjects with respect to healthy ones under one (T(22) = 2.434987, $\rho = 0.011729$) and two tailed testing hypotheses (T(22) = 2.434987, $\rho = 0.023458$);
(d) CR1, the clause frequency is significantly lower ($\alpha < 0.05$) for depressed subjects with respect to healthy ones under one (T(22) = 4.028679, $\rho = 0.000281$) and two tailed testing hypotheses (T(22) = 4.028679, $\rho = 0.000562$);

Figure 1 illustrates the differences between depressed and healthy subjects in CR1 frequency rate and CR2, DPR1, EPR2 duration rates. Differences were not significant for the remaining measurements.

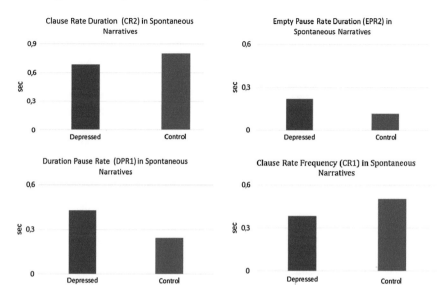

Fig. 1 Differences between depressed and healthy subjects in CR1, CR2, DPR1, EPR2 rates

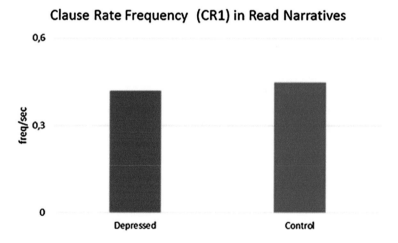

Fig. 2 Differences between depressed and healthy subjects in CR1 frequency rates

In the "read narratives" only the frequency of clauses (CR1) was significantly lower ($\alpha < 0.05$) for the depressed subjects with respect to healthy ones, under one tailed (T(22) = 1.834149, $\rho = 0.040778$) testing hypothesis. Figure 2 illustrates the CR1 differences between the two groups.

4 Discussions

This research investigates on how depressive disorders affect speech pausing strategies either in spontaneous or read narratives. In literature, the identified acoustic differences between depressed and healthy speech are attributed to changes in F0 frequency values and F0 related measures, as well as, formants, power spectral density, Mel Frequency Cepstral (MFCC) coefficients, speech rate and glottal parameters such as jitter and shimmer [2, 5, 13, 21, 22, 26, 27, 38]. In particular, it has been proven that depressed speech exhibits a "slow" auditory dimension [12, 25] and it is perceived as sluggish. According to our results, the perceived "sluggishness" can be primarily attributed to lengthened empty pauses (no significant differences were found for filled pause and consonant/vowel lengthening), and shortened phonation time (clauses are shorter in duration and less numerous). Similar results were found in "automatic" speech tasks (such as counting, reciting the alphabet, and/or reading (see [1, 6, 27, 35]) but not in "extemporaneous free speech samples" (see [27], p. 9). This different outcome can be attributed to the fact that in [27] the speech was collected through telephone and the noise on the communication line may have distorted the signal making difficult to automatically identify the pause boundaries. In addition, the production of spontaneous speech is more cognitively demanding with respect to automatic speech tasks, requiring preparation, word selection and motor articulatory control. Since speech pauses signal complex cognitive planning processes related to the effort to communicate new information content [10, 18], the lengthened silences and shorter phonation time can be considered an index of cognitive and psychomotor retardations related to the degree of severity of depressive states [8, 25, 31, 34, 36]. Longer silences in depressed speech suggest planning communicative efforts to maintain a regular conversation and the need for more cognitive elaboration time [25, 31, 36]. The extra time requirements may produce inadequacies in a typical interactional exchange, both for the listener that does not know how to react to these long silences and the speaker who feels the failure of the exchange. The requirements for extra time to express themselves reduce the phonation time, reducing the amount of information content to be communicated. The consequence is a social impairment that isolates these subjects and impairs the quality of their social life.

Robust speech features for detecting depressive states might help in the implementation of automatic tools to support doctors in the early detection and diagnosis of depression. Speech silences and phonation time seem robust enough to be automatically exploited, given the significant differences measured in different speech tasks and contexts. An early detection of depression can help patients reduce the length of the depressive period and allow a faster recovery.

The strength of the reported results is that they are in agreement with data obtained through different experimental set-ups, supporting the robustness of speech pauses as feature based algorithms. However, there is the need for more data and more thorough analyses to exploit them as depression diagnosis support.

Acknowledgments This work contributes to the research developed in the European Space Agency (ESA) project: Psychological Status Monitoring by Computerised Analysis of Language phenomena (COALA, AO-11-Concordia).

References

1. Alpert, M., Pouget, E.R., Silva, R.R.: Reflection of depression in acoustic measures of the patient's speech. J. Affect. Disord. **66**, 59–69 (2001)
2. Alghowinem, S., et al.: Detecting depression: a comparison between spontaneous and read speech. In: IEEE 3 International Conference on Acoustics, Speech, and Signal Processing (ICASSP), pp. 7547–7551 (2013)
3. Beck, A.T., Steer, R.A., Brown, G.K.: Manual for the Beck Depression Inventory-II. Psychological Corporation, San Antonio, TX (1996)
4. Benus, S., Enos, F., Hirschberg, J., Shriberg, E.: Pauses in Deceptive Speech, Prosody 2006, Dresden. http://www1.cs.columbia.edu/nlp/papers/2006/benus_al_06.pdf (2010)
5. Boonla, T., Yingthawornsuk, T.: Assessment of vocal correlates of clinical depression in female subjects with probabilistic mixture modeling of speech cepstrum. In: IEEE 11th International Conference on Control, Automation and Systems (ICCAS 2011), 26–29 Oct 2011
6. Bouhuys, A.L., et al.: Speech timing measures of severity, psychomotor retardation, and agitation in endogenously depressed patients. J. Common Disord. **17**, 277–288 (1984)
7. Butterworth, B.L.: Evidence for pauses in speech. In: Butterworth, B.L. (ed.) Language Production: Speech and Talk, vol. 1, pp. 155–176. London Academic Press (1980)
8. Caligiuri, M.P., Ellwanger, J.: Motor and cognitive aspects of motor retardation in depression. J. Affect. Disord. **57**, 83–93 (2000)
9. Campine, E., Véronis, J.: Pauses and hesitations in french spontaneous speech. In: Disfuency in Spontaneous Speech. Workshop in International Speech Communication Association, pp. 43–46 (2005)
10. Chafe, W.: Cognitive constraint on information flow. In: Tomlin, R. (ed.) Coherence and Grounding in Discourse John Benjamins, pp. 20–51 (1987)
11. Clark, H.H., Fox Tree, J.E.: Using uh and um in spontaneous speaking. Cognition **84**, 73–111 (2002). Goldman-Eisler, F.: Sequential temporal patterns and cognitive processes in speech. Lang. Speech **10**(2), 122 (1967)
12. Cohn, J.F., et al.: Detecting Depression from Facial Actions and Vocal Prosody. In: 3rd International Conference on Affective Computing and Intelligent Interaction (ACII 2009) (2009)
13. Daniel, J., et al.: Acoustical properties of speech as indicators of depression and suicidal risk. IEEE Trans. Biomed. Eng. **47**(7), 829–837 (2000)
14. Esposito, A.: On vowel height and consonantal voicing effects: data from Italian. Phonetica **59**(4), 197–231 (2002)
15. Esposito, A.: Pausing strategies in children. In: Proceedings of NOLISP05, International Conference in Nonlinear Speech Processing, pp. 42–48. Cargraphics, Barcelona, Spain, 19–22 April 2005
16. Esposito, A.: Children's organization of discourse structure through pausing means. In: Faundez-Zanuy, M., et al. (ed.), Nonlinear Analyses and Algorithms for Speech Processing, Lecture Notes in Computer Science (LCNS), vol. 3817, pp. 108–115, Springer-Verlag, Berlin Heidelberg (2006)
17. Esposito, A., Stejskal, V., Smékal, Z., Bourbakis, N.: The significance of empty speech pauses: cognitive and algorithmic issues. In: Mele, F., et al. (eds.) Brain Vision Artificial Intelligence, LCNS, vol. 4729, pp. 542–554. Springer, Berlin Heidelberg (2007)
18. Esposito, A., Marinaro, M.: What pauses can tell us about speech and gesture partnership. In: Esposito, A., Bratanic, M., Keller, E., Marinaro, M. (Eds.), Fundamentals of Verbal and Nonverbal Communication and the Biometric Issue, NATO Publishing Series, Sub-Series E: Human and Societal Dynamics, vol. 18, pp. 45–57. IOS press, The Netherlands (2007)

19. Esposito, A., Esposito, A.M.: On speech and gesture synchrony. In: Esposito, A. et al. (Eds), Analysis of Verbal and Nonverbal Communication and Enactment: The Processing Issue. LNCS 6800, pp. 252–272. ISBN 978-3-642-25774-2, Springer-Verlag, Berlin Heidelberg (2011)
20. Ghisi, M., Flebus, G.B., Montano, A., Sanavio, E., Sica, C.: Beck Depression Inventory-II. Manuale italiano. Organizzazioni Speciali, Firenze (2006)
21. Helfer, B.S., et al.: Classification of depression state based on articulatory precision. In: 14th Annual Conference of the International Speech Communication Association, pp. 1–5 (2013)
22. Hewlett Sanchez, M., et al. Using prosodic and spectral parameters in detecting depression in elderly males. In: INTERSPEECH 2011, pp. 3001–3004. Florence, Italy, 27–31 Aug 2011
23. Maclay, H., Osgood, C.E.: Hesitation phenomena in spontaneous english speech. Word **15**, 19–44 (1959)
24. MacNeilage, P.: The motor control of the serial ordering of the speech. Psychol. Ref. **77**(3), 182 (1970)
25. Marazziti, D., Consoli, G., Picchetti, M., Carlini, M., Faravelli, L.: Cognitive impairment in major depression. Eur. J. Pharmacol. **626**, 83–86 (2010)
26. Moore, E., Clements, M., Peifer, J., Weisser, L.: Investigating the role of glottal parameters in classifying clinical depression. In: Proceedings of the 25th Annual International Conference of the IEEE Engineering in Medicine and Biology Society IEEE 25th, vol. 3, pp. 2849–2852 (2003)
27. Mundt, J.C., Snyder, P.J., Cannizzaro, M.S., Chappie, K., Geralts, D.S.: Voice acoustic measures of depression severity and treatment response collected via interactive voice response (IVR) technology. J. Neurolinguistics **20**, 50–64 (2007)
28. Oliveira, M.: Prosodic Features in Spontaneous Narratives, Ph.D. Thesis, Simon Fraser University (2000)
29. Oliveira, M.: Pausing strategies as means of information processing narratives. In: Proceeding of the International Conference on Speech Prosody, Ain-en-Provence, pp. 539–542 (2002)
30. O'Shaughnessy, D.: Timing patterns in fluent and disfluent spontaneous speech. In: Proceedings of ICASSP Conference, pp. 600–603. Detroit (1995)
31. Porter, R.J., Gallagher, P., Thompson, J.M., Young, A.H.: Neurocognitive impairment in drug free patients with major depressive disorder. Br. J. Psychiatry **182**, 214–220 (2003)
32. Richmond, V., Payne, McCroskey J., Payne, S.: Nonverbal Behavior in Interpersonal Relations. Prentice Hall, Upper Saddle River (1991)
33. Rochester, S.R.: The significance of pauses in spontaneous speech. J. Psycholinguist. Res. **2**(1), 51 (1973)
34. Sabbe, B., et al.: Retardation in depression: assessment by means of simple motor tasks. J. Affect. Disord. **55**, 39–44 (1999)
35. Szabadi, E., Bradshaw, C.M., Besson, A.O.: Elongation of pause-time in speech: a simple, objective measure of motor retardation in depression. Br. J. psychiatry **129**, 7–592 (1976)
36. Tsourtos, G., Thompson, J.C., Stough, C.: Evidence of an early information processing speed deficit in unipolar major depression. Psychol. Med. **32**, 259–265 (2002)
37. Vinciarelli, A., Chatziioannou, P., Esposito, A.: When the words are not everything: the use of laughter, fillers, back-channel, silence, and overlapping speech in phone calls. Frontiers in ICT, vol. 4, article 4, pp. 1–11 (2015). doi:10.3389/fict.2015.00004
38. Yingthawornsuk, T., et al.: Characterizing sub-band spectral entropy based acoustics as assessment of vocal correlate of depression. In: IEEE 10th International Conference on Control Automation and Systems (ICCAS 2010), p. 1183, 27–30 Oct 2010

Perceptual Features as Markers of Parkinson's Disease: The Issue of Clinical Interpretability

Jiri Mekyska, Zdenek Smekal, Zoltan Galaz, Zdenek Mzourek, Irena Rektorova, Marcos Faundez-Zanuy and Karmele López-de-Ipiña

Abstract Up to 90 % of patients with Parkinson's disease (PD) suffer from hypokinetic dysathria (HD) which is also manifested in the field of phonation. Clinical signs of HD like monoloudness, monopitch or hoarse voice are usually quantified by conventional clinical interpretable features (jitter, shimmer, harmonic-to-noise ratio, etc.). This paper provides large and robust insight into perceptual analysis of 5 Czech vowels of 84 PD patients and proves that despite the clinical inexplicability the perceptual features outperform the conventional ones, especially in terms of discrimination power (classification accuracy ACC = 92 %, sensitivity SEN = 93 %, specificity SPE = 92 %) and partial correlation with clinical scores like UPDRS (Unified Parkinson's disease rating scale), MMSE (Mini-mental state examination) or FOG (Freezing of gait questionnaire), where $p < 0.0001$.

Keywords Perceptual features · Perceptual analysis · Parkinson's disease · Hypokinetic dysarthria · Speech processing

J. Mekyska (✉) · Z. Smekal · Z. Galaz · Z. Mzourek
Department of Telecommunications, Brno University of Technology,
Technicka 10, 61600 Brno, Czech Republic
e-mail: mekyska@feec.vutbr.cz

I. Rektorova
First Department of Neurology, St. Anne's University Hospital,
Pekarska 53, 65691 Brno, Czech Republic

I. Rektorova
Applied Neuroscience Research Group, Central European Institute of Technology,
Masaryk University, Komenskeho Nam. 2, 60200 Brno, Czech Republic

M. Faundez-Zanuy
Escola Universitaria Politecnica de Mataro, Tecnocampus,
Avda. Ernest Lluch 32, 08302 Mataro, Barcelona, Spain

K. López-de-Ipiña
Department of Systems Engineering and Automation, University of the Basque Country
UPV/EHU, Av de Tolosa 54, 20018 Donostia, Spain

© Springer International Publishing Switzerland 2016
A. Esposito et al. (eds.), *Recent Advances in Nonlinear Speech Processing*,
Smart Innovation, Systems and Technologies 48,
DOI 10.1007/978-3-319-28109-4_9

1 Introduction

Parkinson's disease (PD) is a neurodegenerative disease caused by a progressive loss of dopaminergic neurons, primarily in the substantia nigra pars compacta, but also in other parts of brain [11]. Prevalence of this disease is estimated to 1.5 % for people aged over 65 years [9]. PD is associated with different motor and non-motor deficits like muscular rigidity, rest tremor, bradykinesia and postural instability [3, 6, 11]. In 60–90 % of PD patients the multimodal disruption of motor speech realization called hypokinetic dysarthria (HD) can be observed [4]. Most of patients with HD have soft and breathy voice with small variation in speech intensity (monoloudness) and fundamental frequency (monopitch) [1]. The other clinical signs like decreased artic-ulatory organs movement, hoarse or harsh voice, flat speech melody (dysprosody) or voice tremor can be observed as well [10].

In the last two decades scientists developed several acoustic signal analysis meth-ods focused on assessment of parkinsonic speech [5, 8, 13]. Although a lot has already been investigated, some issues (e.g. early stage detection or accurate progress esti-mation) have not been solved yet. As time goes, new, robust and more sophisticated speech parametrization methods occur. But this speech features evolution more often builds barrier between engineers and clinicians, which is called "The issue of clinical interpretability". A feature with high discrimination power or good abilities to mon-itor progress of disease can be proposed, however it is becoming useless as soon as we try to find relations between its value and clinical signs of HD. In order to make a good diagnose the clinicians need transparent parametrization. In other words, when a value of feature changes they must know what will be the result from the clini-cal sign point of view. According to this consideration we can divide features into two categories: (1) clinically interpretable—they help us to directly quantify clini-cal signs; (2) clinically inexplicable features—we can find significant correlations between their values and clinical signs, but we can just guess what are the exact relations.

Perceptual features are good representatives of the second category. Although some researchers tried to interpret them from hypokinetic dysarthria signs point of view [2, 7, 13], their meaning is still hidden in this field of science. Probably the deepest research focused on discrimination power of perceptual features was made by Orozco-Arroyave et al. [7]. Their results show that perceptual analysis of sustained Spanish vowels [a], [i] and [u] based on PLP (Perceptual Linear Predictive Coefficients) or MFCC (Mel-Frequency Cepstral Coefficients) provides the highest discrimination power. However, they used just a limited set of features (5) and small group of patients and control speakers respectively (20 + 20).

To sum up the introduction, although the perceptual features are clinically inex-plicable, they could be very good markers of Parkinson's disease. Therefore the aim of this work is to: (1) prove that perceptual features can outperform the conventional clinically interpretable parameters or significantly improve PD identification accu-racy; (2) test a large set of perceptual parameters and identify feature with the highest discrimination power; (3) find what kind of vowel realization it is better to analyse; (4) identify perceptual features that can predict values of different clinical tests.

The rest of this paper is organized as follows. Sections 2 and 3 describe the dataset and methodology respectively. Section 4 provides some preliminary results where the features are evaluated in terms of correlation and mutual information with speakers' label. Results of single-feature classification are given as well. Finally partial correlation with clinical tests and classification based on feature selection is considered. The conclusion is given in Sect. 5.

2 Data

In the frame of this study 84 PD patients (36 women, 48 men) and 49 (24 women, 25 men) age and gender matched healthy controls (HC) were enrolled at the First Department of Neurology, St. Anne's University Hospital in Brno, Czech Republic. The demographic and clinical characteristics of PD patients can be seen in Table 1. The healthy controls had no history or presence of brain diseases (including neurological and psychiatric illnesses) or speech disorders. The PD patients were on their regular dopaminergic treatment. All participants signed an informed consent form that had been approved by the Ethics Committee of St. Anne's University Hospital in Brno.

During acquisition the participants were asked to utter sequence of 5 Czech vowels ([a], [e], [i], [o] and [u]) in 4 different ways: (1) s—short vowels pronounced with normal intensity; (2) l—sustained vowels pronounced with normal intensity; (3) ll—sustained vowels pronounced with maximum intensity; (4) ls—sustained vowels

Table 1 Demographic and clinical characteristics of PD patients

Speakers	PD (females)	PD (males)
Number	36	48
Age (years)	68.47 ± 7.64	66.21 ± 8.78
PD duration (years)	7.61 ± 4.85	7.83 ± 4.39
UPDRS III	22.06 ± 13.73	26.85 ± 10.22
UPDRS IV	2.72 ± 3.01	3.15 ± 2.59
RBDSQ	3.42 ± 3.48	3.85 ± 2.99
FOG	6.94 ± 5.72	6.67 ± 5.57
NMSS	36.03 ± 26.72	38.19 ± 19.72
BDI	18.57 ± 23.94	9.69 ± 6.23
MMSE	27.38 ± 3.63	28.56 ± 1.05
LED (mg)	862.44 ± 508.3	1087 ± 557.47

[a] UPDRS III—Unified Parkinson's disease rating scale, part III: Motor Examination; UPDRS IV—Unified Parkinson's disease rating scale, part IV: Complications of Therapy; RBDSQ—The REM sleep behavior disorder screening questionnaire); FOG—Freezing of gait questionnaire; NMSS—Non-motor symptoms scale; BDI—Beck depression inventory; MMSE—Mini-mental state examination; LED—L-dopa equivalent daily dose

pronounced with minimum intensity, but not whispered. Speech was recorded with sampling frequency $f_s = 48\,\text{kHz}$ and consequently downsampled to $f_s = 16\,\text{kHz}$ in order to decrease computational burden.

3 Methodology

In order to compare the discrimination power of perceptual features to conventional ones, during the parametrization step we extracted fundamental frequency F_0, 5 kinds of jitter and 6 kinds of shimmer, Teager-Kaiser energy operator TKEO, formants $F_1 - F_3$ and their bandwidths $BW_1 - BW_3$, harmonic-to-noise ratio HNR, glottal-to-noise excitation ratio GNE, vowel space area VSA and its logarithmic version lnVSA, formant centralization ratio FCR, vowel articulation index VAI and ratio of second formants of vowels [i] and [u] F_{2i}/F_{2u}. If the specific feature was represented by vector or matrix, we applied transformation to scalar value. For this purpose we used median, standard deviation (std), 1st percentile (1p), 99th percentile (99p) and interpercentile range (ir) defined as 99p–1p. In the case of matrix the transformation was applied over each band separately.

3.1 *Perceptual Features*

First of all we included to this study the most popular MFCC (Mel Frequency Cepstral Coefficients) that can indirectly detect slight misplacements of articulators [13]. Consequently we derived from MFFC next 3 kinds of perceptual features: LFCC (Linear Frequency Cepstral Coefficients), CMS (Cepstral Mean Subtraction coefficients) and MFCC adjusted to equal loudness curves (as in the case of PLP). In order to provide information complementary to MFCC we tested MSC (Modulation Spectra Coefficients).

Next set of features is based on linear prediction: LPC (Linear Predictive Coefficients), PLP (Perceptual Linear Predictive coefficients), LPCC (Linear Predictive Cepstral Coefficients), LPCT (Linear Predictive Cosine Transform coefficients) and ACW (Adaptive Component Weighted coefficients). In comparison to simple LPC or MFCC the PLP also takes into account adjustment to the equal loudness curve and intensity-loudness power law. The advantage of LPCC and LPCT over "classic" LPC is small correlation of values. Finaly the advantage of ACW is that these coefficients are less sensitive to channel distortion.

The last features in this study are ICC (Inferior Colliculus Coefficients) that analyse amplitude modulations in voice using a biologically-inspired model of the inferior colliculus. All perceptual features were extended by their 1st order regression coefficients (Δ).

3.2 Preliminary Analysis

We employed calculation of Spearman's rank sum correlation and mutual information (MI) between feature vectors and resulting speakers' label in order to estimate discrimination power of the vowels separately. Consequently we applied Mann-Whitney U test and classification based on random forests (RF). Classification results were expressed by ACC, SEN, SPE and trade-off between sensitivity and specificity (TSS) defined as:

$$TSS = 2^{\sin\left(\frac{\pi \cdot SEN}{2}\right)\sin\left(\frac{\pi \cdot SPE}{2}\right)} \tag{1}$$

Finally to identify perceptual features that can predict values of different clinical tests we calculated Spearman's partial correlations where the effect of patients' age and L-dopa equivalent daily dose was removed.

3.3 Classification

In the last step we performed classification with a two-step feature selection. Firstly we reduced set of features to 500 parameters by mRMR (minimum Redundancy Maximum Relevance) and consequently we employed SFFS (Sequential Forward Feature Selection). Three scenarios were considered: individual vowel analysis; classification within each vowel sequence (see Sect. 2); classification using all vowel realizations. In all the cases we used leave-one-out validation.

4 Experimental Results

The preliminary results performed by Spearman's rank correlation, mutual information, Mann-Whitney U test and RF classifier are given in Table 2. The results of PD identification based on feature selection can be seen in Table 3. Finally the results of Spearman's partial correlations between clinical characteristics and selected features are in Table 4.

According to the preliminary analysis we can conclude that std of 10th CMS coefficient extracted from short vowel [o] provides the best discrimination power in terms of ACC (80.45 %), SEN (78.57 %), SPE (83.67 %) and TSS (1.88). On the other hand conventional shimmer extracted from sustained vowel [i] pronounced with minimum intensity reached better results of $\rho(-0.4064)$, MI (0.7633) and p (0.0000).

Considering the classification using feature selection, in the first scenario (individual vowel analysis) we can observe the best results in the case of sustained and loudly pronounced vowel [a] (ACC = 91.73 %, SEN = 90.48 %, SPE = 93.88 %, TSS = 1.98). All 8 selected features were perceptual. In the case of second scenario (classification within each vowel sequence) the best results provided sustained vowels

Table 2 Individual vowel analysis

Vowel	Feature	ρ	MI	p	ACC (%)	SEN (%)	SPE (%)	TSS
a (s)	10th ICC (99p)	−0.1356	0.0784	0.1198	71.43	72.62	69.39	1.75
e (s)	7th LPCC (99p)	−0.0057	0.6975	0.9498	69.92	72.62	65.31	1.71
i (s)	17th ΔMFCC (1p)	0.1242	0.7349	0.1542	69.92	70.24	69.39	1.73
o (s)	10th CMS (std)	0.0059	0.0611	0.9477	**80.45**	**78.57**	**83.67**	**1.88**
u (s)	6th ACW (1p)	0.1003	0.7296	0.2502	72.18	75.00	67.35	1.75
a (l)	9th LFCC (std)	−0.2042	**0.7963**	0.0191	75.19	75.00	75.51	1.81
e (l)	15th MSC (1p)	0.0938	0.0050	0.2823	66.92	65.48	69.39	1.69
i (l)	14th ICC (1p)	−0.0503	0.0845	0.5646	68.42	69.05	67.35	1.71
o (l)	2nd ΔLPCT (median)	0.0978	0.7069	0.2619	73.68	77.38	67.35	1.76
u (l)	12th CMS (std)	−0.0842	0.0269	0.3347	77.44	73.81	83.67	1.85
a (ll)	18th ΔLPCC (ir)	0.0958	0.7576	0.2720	72.18	72.62	71.43	1.76
e (ll)	11th ΔPLP (99p)	0.2635	0.6129	0.0025	72.18	75.00	67.35	1.75
i (ll)	5th PLP (std)	−0.0321	0.6460	0.7142	68.42	66.67	71.43	1.72
o (ll)	10th ΔCMS (ir)	−0.2038	0.7643	0.0193	75.94	73.81	79.59	1.83
u (ll)	17th CMS (std)	−0.0486	0.0325	0.5783	72.18	76.19	65.31	1.74
a (ls)	13th ACW (ir)	−0.0093	0.7199	0.9164	73.68	79.76	63.27	1.74
e (ls)	8th CMS (std)	0.1887	0.0835	0.0304	72.18	64.29	85.71	1.77
i (ls)	shimmer (local.dB)	**−0.4064**	0.7633	**0.0000**	72.18	75.00	67.35	1.75
o (ls)	3rd ICC (99p)	0.1324	0.0325	0.1289	69.17	71.43	65.31	1.71
u (ls)	9th CMS (std)	−0.0191	0.0232	0.8282	75.19	69.05	85.71	1.82

[a] ρ—Spearman's rank correlation coefficient; MI—mutual information; p—significance level (Mann-Whitney U test; ACC—classification accuracy; SEN—sensitivity; SPE—specificity; TSS—trade-off between sensitivity and specificity; s—short vowel pronounced with normal intensity; l—sustained vowel pronounced with normal intensity; ll—sustained vowel pronounced with maximum intensity; ls—sustained vowel pronounced with minimum intensity (not whispering)

Table 3 Classification results (using feature selection)

Vowels	ACC (%)	SEN (%)	SPE (%)	TSS	No.
a (s)	84.21	86.90	79.59	1.90	6
e (s)	81.95	82.14	81.63	1.89	8
i (s)	72.18	73.81	69.39	1.76	3
o (s)	80.45	78.57	83.67	1.88	1
u (s)	85.71	86.90	83.67	1.93	6
a (l)	87.22	88.10	85.71	1.94	6
e (l)	75.94	78.57	71.43	1.80	7
i (l)	82.71	83.33	81.63	1.90	11
o (l)	75.19	79.76	67.35	1.77	2
u (l)	77.44	73.81	83.67	1.85	1
a (ll)	**91.73**	**90.48**	**93.88**	**1.98**	8
e (ll)	78.20	83.33	69.39	1.81	3
i (ll)	78.95	82.14	73.47	1.84	6
o (ll)	81.20	80.95	81.63	1.89	3
u (ll)	72.18	76.19	65.31	1.74	1
a (ls)	76.69	78.57	73.47	1.82	3
e (ls)	87.97	88.10	87.76	1.95	8
i (ls)	84.21	84.52	83.67	1.92	11
o (ls)	76.69	77.38	75.51	1.83	6
u (ls)	84.21	86.90	79.59	1.90	4
all (s)	80.45	78.57	83.67	1.88	1
all (l)	**91.73**	**90.48**	**93.88**	**1.98**	9
all (ll)	81.95	79.76	85.71	1.90	7
all (ls)	90.98	91.67	89.80	1.97	11
all (s, l, ll, ls)	**92.48**	**92.86**	**91.84**	**1.98**	9

[a] ACC—classification accuracy; SEN—sensitivity; SPE—specificity; TSS—trade-off between sensitivity and specificity; No.—number of selected features, s—short vowel pronounced with normal intensity; l—sustained vowel pronounced with normal intensity; ll—sustained vowel pronounced with maximum intensity; ls—sustained vowel pronounced with minimum intensity (not whispering)

pronounced with natural intensity (ACC = 91.73 %, SEN = 90.48 %, SPE = 93.88 %, TSS = 1.98), where all 9 selected features were perceptual as well. It was proved that in order to get best classification results (ACC = 92.48 %, SEN = 92.86 %, SPE = 91.84 %, TSS = 1.98) it is advantageous to use all 4 sets of vowels.

In our recent study we found out that sustained vowels pronounced with minimum intensity can be good speech tasks for detection of improper vocal folds vibration (measured by features based on empirical mode decomposition) [12]. In the case of perceptual analysis we observe that loudly pronounced features are better candidates to analyse. We explain this by substantiality of perception. Theoretically longer and more intense stimuli results in better perception.

Table 4 Spearman's partial correlations between clinical characteristics and selected features (after removal of age and LED effect)

Clinical info	Feature	ρ	p
PD duration	i (l): 15th CMS (std)	−0.4369	$3.25 \cdot 10^{-5}$
UPDRS III	i (l): 1st ΔPLP (1p)	−0.5174	$6.98 \cdot 10^{-7}$
UPDRS IV	e (ll): 5th ΔMFCC (ir)	0.4572	$1.23 \cdot 10^{-5}$
RBDSQ	u (ls): 13th ΔMFCC (99p)	0.4906	$2.16 \cdot 10^{-6}$
FOG	a (ls): 6th MFCC (std)	−0.4476	$1.96 \cdot 10^{-5}$
NMSS	a (ll): 12th LPC (99p)	0.4616	$1.25 \cdot 10^{-5}$
BDI	u (s): 3rd ΔLPCT (1p)	0.5832	$1.25 \cdot 10^{-6}$
MMSE	i (l): 20th MFCC (99p)	−0.4719	$5.55 \cdot 10^{-5}$

Finally we have proved that perceptual features significantly correlate ($p < 0.0001$) with different clinical information like UPDRS III (Unified Parkinson's disease rating scale, part III: Motor Examination), UPDRS IV (part IV: Complications of Therapy), RBDSQ (The REM sleep behavior disorder screening questionnaire), FOG (Freezing of gait questionnaire), NMSS (Non-motor symptoms scale), BDI (Beck depression inventory) and MMSE (Mini-mental state examination). This means that they can be used for estimation of these scores.

5 Conclusion

In this paper we perceptualy analysed phonation of 84 PD patients and 49 gender and age matched controls. We achieved all goals of this work: (1) We have proved that perceptual features outperform the conventional ones in terms of discrimination power. (2) From a wide range of perceptual features we have found out that those based on CMS (derived from MFCC) better quantify the signs of hypokinetic dysarthria. (3) We have shown that it is advantageous to perform perceptual analysis of loud sustained vowels. (4) In the case of each considered clinical score we identified a perceptual feature that can be used for its estimation.

In the near future we would like to move further, perceptualy analyse another speech tasks (spontaneous speech, read sentences, etc.) and focus on each gender individually.

Acknowledgments Research described in this paper was financed by the National Sustainability Program under grant LO1401 and by projects NT13499 (Speech, its impairment and cognitive performance in Parkinson's disease), COST IC1206, project "CEITEC, Central European Institute of Technology": (CZ.1.05/1.1.00/02.0068), FEDER and Ministerio de Economía y Competitividad TEC2012-38630-C04-03.

References

1. Arnold, C., Gehrig, J., Gispert, S., Seifried, C., Kell, C.A.: Pathomechanisms and compensatory efforts related to Parkinsonian speech. Neuroimage Clin **4**, 82–97 (2014)
2. Bocklet, T., Noth, E., Stemmer, G., Ruzickova, H., Rusz, J.: Detection of persons with Parkinson's disease by acoustic, vocal, and prosodic analysis. In: 2011 IEEE Workshop on Automatic Speech Recognition and Understanding (ASRU), pp. 478–483 (2011)
3. Brodal, P.: The Central Nervous System: Structure and Function, 3rd edn. Oxford University Press, Oxford (2003)
4. Chenausky, K., MacAuslan, J., Goldhor, R.: Acoustic analysis of pd speech. Parkinson's Dis. **2011**, 1–13 (2011)
5. Eliasova, I., Mekyska, J., Kostalova, M., Marecek, R., Smekal, Z., Rektorova, I.: Acoustic evaluation of short-term effects of repetitive transcranial magnetic stimulation on motor aspects of speech in Parkinson's disease. J. Neural Transm. **120**(4), 597–605 (2013)
6. Mekyska, J., Smekal, Z., Kostalova, M., Mrackova, M., Skutilova, S., Rektorova, I.: Motor aspects of speech imparment in Parkinson's disease and their assessment. Cesk. Slov. Neurol. N. **74**(6), 662–668 (2011)
7. Orozco-Arroyave, J., Arias-Londono, J., Vargas-Bonilla, J., Noth, E.: Perceptual analysis of speech signals from people with Parkinson's disease. Natural and Artificial Models in Computation and Biology. Lecture Notes in Computer Science, vol. 7930, pp. 201–211. Springer, Berlin Heidelberg (2013)
8. Rusz, J., Cmejla, R., Ruzickova, H., Ruzicka, E.: Quantitative acoustic measurements for characterization of speech and voice disorders in early untreated Parkinson's disease. J. Acoust. Soc. Am. **129**(1), 350–367 (2011)
9. Sapir, S., Ramig, L., Fox, C.: Speech and swallowing disorders in Parkinson disease. Curr. Opin. Otolaryngol. Head Neck Surg. **16**(3), 205–210 (2008)
10. Skodda, S., Grnheit, W., Mancinelli, N., Schlegel, U.: Progression of voice and speech impairment in the course of Parkinson's disease: a longitudinal study. Parkinson's Dis. **2013**, 1–8 (2013)
11. Skodda, S., Visser, W., Schlegel, U.: Short- and long-term dopaminergic effects on dysarthria in early Parkinson's disease. J. Neural Transm. **117**, 197–205 (2010)
12. Smekal, Z., Mekyska, J., Galaz, Z., Mzourek, Z., Rektorova, I., Faundez-Zanuy, M.: Analysis of phonation in patients with Parkinson's disease using empirical mode decomposition. In: 2015 International Symposium on Signals, Circuits and Systems (ISSCS), pp. 1–4 (2015)
13. Tsanas, A., Little, M.A., McSharry, P.E., Ramig, L.O.: Nonlinear speech analysis algorithms mapped to a standard metric achieve clinically useful quantification of average Parkinson's disease symptom severity. J. R. Soc. Interface **8**(59), 842–855 (2010)

Phonation Biomechanics in Quantifying Parkinson's Disease Symptom Severity

P. Gómez-Vilda, A. Álvarez-Marquina, A. Tsanas, C.A. Lázaro-Carrascosa, V. Rodellar-Biarge, V. Nieto-Lluis and R. Martínez-Olalla

Abstract It is known that Parkinson's Disease (PD) leaves marks in phonation dystonia and tremor. These marks can be expressed as a function of biomechanical characteristics monitoring vocal fold tension and imbalance. These features may assist tracing the neuromotor activity of laryngeal pathways. Therefore these features may be used in grading the stage of a PD patient efficiently, frequently and remotely by telephone or VoIP channels. The present work is devoted to describe and compare the PD symptom severity quantification from neuromotor-sensitive features with respect to other features on a telephone-recorded database. The results of these comparisons are presented and discussed.

Keywords Neurologic disease · Parkinson's Disease (PD) · Speech neuromotor activity · Aging voice · Dysarthria

1 Introduction

Parkinson's Disease (PD) is a neurodegenerative disorder occurring due to deterioration of *substantia nigra* in midbrain, with increasing yearly prevalence and incidence rates. Its prevalence is expected to double in 2030 with respect to 2005 [1]. It is well known that PD affects voice and speech even at an early stage, when other symptoms are not yet evident [2, 3]. Therefore speech features have been routinely used to detect, assess and monitor PD by clinicians [4, 5]. The Unified PD Rating Scale (UPDRS) [4, 6, 7] is often used in PD clinical evaluation, assigning a normalized

P. Gómez-Vilda (✉) · A. Álvarez-Marquina · C.A. Lázaro-Carrascosa · V. Rodellar-Biarge ·
V. Nieto-Lluis · R. Martínez-Olalla
Neuromorphic Speech Processing Lab, Center for Biomedical Technology,
Universidad Politécnica de Madrid, Campus de Montegancedo, Pozuelo de Alarcón,
28223 Madrid, Spain
e-mail: pedro@fi.upm.es

A. Tsanas
Oxford Centre for Industrial and Applied Mathematics, University of Oxford,
Oxford, UK

© Springer International Publishing Switzerland 2016
A. Esposito et al. (eds.), *Recent Advances in Nonlinear Speech Processing*,
Smart Innovation, Systems and Technologies 48,
DOI 10.1007/978-3-319-28109-4_10

score in the interval 0–4 on 17 items such as memory, speech, swallowing, hand-writing, walking and numbness, amongst others. Another popular PD grading scale is that of Hoehn and Yahr (H&Y) [8]. During the last decade, important advances in PD evaluation and grading using speech have been produced [9]. These are based in correlates of phonation, prosody or fluency. Phonation correlates are based in f0 distortion measurements as jitter, shimmer, harmonic-noise-ratios (HNR), pitch period entropy (PPE) or mel-frequency cepstral coefficients (MFCC's), among others [4]. The problem in using these correlates is the loss of semantics, which may imply a too large penalty jeopardizing further research. The present approach is intended to advance in the grading of PD whilst maintaining the semantics of the domain. A set of biomechanical features derived from phonation which are known to preserve parameter semantics [10, 11] have been used. These are related to vocal fold mechanical stress and tremor, which is one of the primary hallmarks that may be perceived in around 60 % of the PD cases, manifesting in bands around 2–4 Hz (physiological), 6–10 (neurological) or above 10 Hz (sometimes addressed as jitter or flutter). If it is concentrated in the upper bands it may not be perceived acoustically most times. The performance of these features will be compared with the f0 distortion features. A brief description of how PD affects the biomechanics of phonation is provided in Sect. 2. Section 3 presents the methodology of the study. Results are shown and discussed in Sect. 4. Conclusions are presented in Sect. 5.

2 Larynx Neuro-Motor Activation Features

The speech neuromotor sequence activates the muscles of the pharynx, tongue, larynx, chest and diaphragm through sub-thalamic pathways. The *cricothyroid, transverse* and *oblique arytenoid*, as well as the *posterior cricoarytenoid* muscles in the larynx are especially relevant in phonation, as they are responsible for vocal fold stretching, adduction and abduction. Fine muscular control is provided by a two-way regulation system in which dopamine, noradrenaline, serotonine and acetylcholine are involved. Alterations in the level of these substances, and especially of dopamine produced in *substantia nigra* impairs the regulatory function and results in the appearing of the PD motor syndrome characterized by tremor, rigidity bradykinesia and loss of equilibrium [12]. Perturbations in respiration, phonation and articulation affect speech. The dystonic behavior of the vocal fold stiffness or its fluctuation (tremor), are some of the symptoms associated to PD phonation. The method proposed here is to estimate vocal fold stiffness from sustained long vowels (/a/). The procedures used in the estimation of this correlate are vocal tract inversion by a lattice adaptive filter [13], and biomechanical inversion of a 2-mass model of the vocal folds [14]. As a result, an estimate of the vocal fold body mechanical stiffness is produced for each phonation cycle, from which tremor may be characterized (see [10] for more details). The mechanical stiffness of the vocal fold body and cover, the dynamic mass of body and cover, the dissipation losses in the oscillating vocal folds, together with the cycle-to-cycle variation of these parameters (asymmetric vibration unbalance), can be used as features. The tremor band-split inverse model coefficients, as well as tremor distribution in frequency and amplitude [11] can also be used as correlates.

3 Materials and Methods

This study is framed within the Patient Voice Analysis Challenge (PVAC) supported by Synapse [15]. The PVAC has recruited a large database of telephone-line quality sustained /a/ utterances recorded at 8 kHz from crowdsourcing volunteers suffering from any degree of PD, who were also asked to self-assess symptom severity under the UPDRS and H&Y standards. The data made available by PVAC is a set of /a/ utterances in .wav format, plus the data sheet with the UPDRS and H&Y scores, and a collection of features already produced for each recording summarized in Table 1.

The purpose of the present study is to compare the performance of larynx neuro-motor activity features (LNMAF) described in Sect. 2 with f0 distortion and MFCC features provided by PVAC (PVACF), which are known to be highly competent in PD detection and grading [4–7]. The set of biomechanical features used in the present study are given in Table 2.

The experiments are designed to correlate the UPDRS scores provided by the patients, with those predicted using feature sets PVACF and LNMAF. The data set including 770 sample utterances was split randomly into training and test subsets, composed of 389 and 390 samples each. A male and a female training subsets were produced selecting 183 and 186 samples from the training subset respectively (one sample per subject). 38 features (as the ones listed in Table 1) were estimated for each sample in the subsets by the PVAC. The training sets were processed using

Table 1 Features provided within PVAC (PVACF)

F001. Median of fundamental frequency f0
F002. Mean absolute f0 time derivative
F003. Median absolute f0 time derivative
F004. Mean of abs. val. of RMS power time derivative
F005. Median of abs. val. of RMS power time derivative
F006–19. Median of MFCC's 0–12 across entire voice recording
F020–32. Mean of MFCC's 0–12 across entire voice recording
F033. Recurrence period density entropy (RPDE) hnorm
F034. Detrended fluctuation analysis (DFA) scaling parameter alpha
F035. Modified pitch period entropy (PPE)
F036. Relative spectral power 0–500 Hz
F037. Relative spectral power 500–1000 Hz
F038. Relative spectral power 1000–2000 Hz

Table 2 Larynx neuro-motor activity features (LNMAF)

F039. Median of fundamental frequency f0
F040. Median of jitter (relative between neighbor phonation cycles)
F041. Median of shimmer (relative between mean amplitude of neighbor phonation cycles)
F042. Median of maximum flow declination rate (MFDR)
F043. Median of noise-harmonic ratio
F044. Median of mucosal wave correlate to average acoustic wave ratio (MWC/AAC)
F045–58. Medians of cepstral coefficients across the analysis window
F059–70. Medians of the mucosal wave correlate power spectral density (MWCPSD) profile
F071–72. Medians of MWCPSD minima slenderness
F073–75. Medians of vocal fold body dynamic mass, losses and stiffness
F076–78. Medians of vocal fold body dynamic mass, losses and stiffness unbalances
F079–81. Medians of vocal fold cover dynamic mass, losses and stiffness
F082–84. Medians of vocal fold cover dynamic mass, losses and stiffness unbalances
F085–86. Medians of glottal source recovery instants 1 and 2
F087–88. Medians of glottal source open instants 1 and 2
F089. Median of glottal source maximum instant
F090–91. Medians of glottal source recovery amplitudes 1 and 2
F092–93. Medians of glottal source open amplitudes 1 and 2
F094–95. Median of glottal flow stop and start instants
F096. Median of glottal flow closing instant
F097–100. Medians of glottal flow gap, contact, adduction and permanent defects
F101–103. Medians of the 1st, 2nd and 3rd order cyclic coefficients (tremor)
F104–105. Medians of the physiological band tremor frequency and amplitude
F106–107. Medians of the neurological band tremor frequency and amplitude
F108–109. Medians of the flutter band tremor frequency and amplitude
F110. Median of the root mean square tremor amplitude

the LNMAF methodology to estimate the 72 features described in Sect. 2 listed in Table 2. The experiments consisted in predicting the subjective self-scores using multiple linear regression Support Vector Machines (SVM's [16]). The goodness of the fit is given in terms of the Weighted Mean Absolute Error (WMAE) defined as:

$$\varepsilon_\omega = \sum_{i=1}^{N} \omega_i \left| s_i - \hat{s}_i \right| \tag{1}$$

where N is the test set size, s_i is the self-score, $\hat{s}_i$ is the score predicted by the SVM, and ω_i is the weight of the prediction (ratio between the number of times subject i appears in the test $-n_i$ contributions-relative to the total number of samples n_i/N). The objective of the experiment is to determine which feature combinations are more efficient in producing the minimal WMAE. The training subsets are divided in 10

subgroups each. For each subgroup a training session of the SVM is conducted, and
a test is carried on the nine other subgroups in a leave-one-out protocol. The scores
obtained are combined in a unified score such as $\hat{s}_i$, and the WMAE is obtained.
The 10-fold validation is repeated on three different feature sets: PVACF, which
includes features F001–38 in Table 1; LNMAF, which includes features F039–110,
and PVACF + LNMAF, which includes the merged feature set F001–110. The combinations of k features producing the lowest WMAE are selected, with k ranging
from 1 to 25. This process is repeated for the three feature sets PVACF, LNMAF and
PVACF + LNMAF. The results are presented and discussed in the next section.

4 Results and Discussion

The feature templates producing the lowest WMAE from the exhaustive run conducted on the k feature templates with $1 < k < 25$ for the three feature sets are
presented in Table 3. The normalized WMAE from male and female sets for the
different feature templates tested is presented in Fig. 1.

It may be seen that the best template features are different for the male and
female subsets, a finding in agreement with previous studies [4]. The WMAE decays
up to a certain point, after which it increases again (a classical manifestation of the
curse of dimensionality). The LNMAF feature set produces lower WMAE's than the
PVACF set for most of the cases. The merging of both feature sets behaves better
than each set alone. The reduction of the WMAE in LNMAF relative to PVACF
is of 16.34 % (males) and 15.22 % (females), and of 25.22 % (males) and 21.76 %
(females) relative to the merged sets. In the PAVFC male set features 12, 15, 16,
19, 21 and 32 correspond to MFCC's; 35 is the modified pitch period entropy of a
logarithmic semitone pitch residual [17]. In the female set features 1 and 3 are f0 and
its time derivative; 5 is the absolute rms value of energy, 8, 12, 13, 16, 18 and 25 are
MFCC's. In the LNMAF best male set features 51 and 52 are unrestricted cepstral
coefficients; 67 is the 2nd maximum frequency position in the glottal source power
spectral density; 73, 74 and 75 are the vocal fold body mass, losses and stiffness; 76
is the vocal fold body mass unbalance (asymmetric vibration); 79 is the vocal fold
cover mass; 85 is the glottal source recovery time; 92 is the glottal source amplitude
at the opening instant; 99 is the vocal fold adduction defect; 105 is the amplitude of
the physiological tremor; and 107 is the amplitude of neurological tremor. Possibly
features 67, 73, 74, 75, 76, 79, 85, 92 and 99 may be affected by aging voice, whereas
105 and 107 show neurodegenerative influence. In the LNMAF best female feature
set 40 is the jitter; 47, 53 and 54 are also unrestricted cepstral coefficients; 61 is
the 2nd maximum frequency amplitude, found in the glottal source power spectral
density; 67 is the 1st maximum frequency position; 74 is the estimate of vocal fold
body losses; 97 and 98 are the glottal flow escape defect, and the contact defect; 101
and 102 are the first two coefficients of the inverse filter estimating of tremor [10],
and 109 is the amplitude of the flutter tremor. When analyzing the features included
in the male set merged templates (PVACF + LNMAF) it may be seen that features 12,

Table 3 Feature templates producing the lowest WMAE (male and female sets)

Set	Temp.	WMAE	Features (males)														
PVACF	8	0.777	12	15	16	19	21	32	35	36	–	–	–	–	–	–	–
LNMAF	13	0.650	51	52	67	73	74	75	76	79	85	92	99	105	107	–	–
PV + LN	11	0.581	12	14	29	38	43	51	57	67	74	99	108	–	–	–	–

Set	Temp.	WMAE	Features (females)														
PVACF	9	0.749	1	3	5	8	12	13	16	18	25	–	–	–	–	–	–
LNMAF	13	0.635	40	47	53	54	61	65	74	97	98	101	102	104	109	–	–
PV + LN	15	0.586	21	22	23	32	39	40	41	47	61	82	83	97	101	102	104

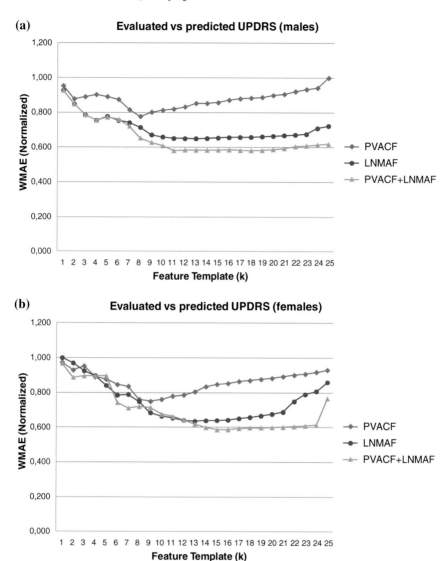

Fig. 1 Normalized WMAE for the different templates tested. Each template index corresponds with the number of features included. **a** Male set. **b** Female set

14 and 29 are MFCC's; 38 is the spectral power in 1–2 kHz; 43 is the noise-harmonic ratio; 51 and 57 are unrestricted cepstral parameters; 67, 74 and 99 are present as well in the LNMAF; and 108 is the flutter frequency. The description of the female set features in the merged templates (PVACF + LNMAF) for the female set is: 21, 22, 23 and 32 are MFCC's; 39 is f0 estimated on the glottal source; 40 and 41 are jitter and shimmer, respectively; 47, 61, 97, 101, 102 and 104 are present as well in the

LNMAF; and 82 and 83 are the unbalances of the vocal fold cover mass and losses. The number LNMAF features in the composite templates (PVACF + LNMAF) is dominant in the male and in the female feature sets when compared with the number of PVACF coefficients (8/3 and 11/4). This fact indicates that the minimum WMAE may be more dependent on larynx activity than on voice distortion features.

5 Conclusions

The severity grading of PD using voice is an intricate, challenging problem. Through the present study a comparative approach to this problem has been presented. The most important findings are the following:

- Glottal source features (LNMAF) are more efficient in reducing WMAE more than voice features (PVACF).
- Combining PVACF and LNMAF results in the minimum WMAE.
- The role of LNMAF features seems to be dominant in the combined feature templates. This result is very relevant, as LNMAF gives information on biomechanical implications which are very semantic.
- These observations are valid for both genders.
- Features related with f0 distortion seem to be highly relevant in all cases. The same is valid for cepstral features, either unrestricted or mel-banded. Biomechanical features (parameters and unbalances) are relevant as well, but it is not clear if they monitor aging voice or PD.
- Contact and opening defects are also relevant, although may be affected by aging.
- Tremor features are very relevant, and are plausibly induced by PD, although the influence of other causes cannot be excluded (essential, spasmodic, etc.).

A word has to be said on the use of telephone-like quality voice, which is justified under the following considerations: database recruiting is based on crowdsourcing, the donators being volunteers providing their own voice on the telephone line, this being a near-real situation if these methods are to be used for distant and frequent evaluation of patients; besides, telephone-quality is subject to channel distortion, therefore if the methodology works well for this type of signal, plausible it will work better for higher quality standards. Finally it must be said that telephone-channel distortion does not affect greatly to biomechanical glottal features, as the spectral power of the glottal signal is concentrated in the 300–3000 Hz, adapting well to channel width. Significant differences between broad and narrow channel band are not noticeable, as demonstrated by the relative invariance to feature statistical distributions with channel width. Furthermore future studies need considering if means are more sensitive to distortion induced by aging or PD voice because of real causes or because of the presence of more outliers in their estimation. Regarding tremor a very important question to deserve further analysis is that of its estimation on voice or on the glottal source. As it is well known, the source-filter model gives a definition to separate the articulation part (acoustic oro-naso-pharyngeal tract) from the glottal source (the glottal flow derivative). But according to the physiological description given in Sect. 2 the innervations of larynx, pharynx, tongue, jaw and facial muscles depend on different neuromotor pathways. Therefore, tremor estimated on voice may be different than that estimated only on the glottal source. To infer the origin and nature of tremor in voice requires a further study on tremor in formants (articulation) and in the glottal source (phonation), this being a pending line of study. Another relevant question is to determine to which extent the features used in grading PD severity are reactive only to PD deterioration or if they are also reactive to aging voice (not necessarily related to PD). Aging voice (presbyphonia) is mainly of organic origin, due in principle to the loss of elastin and collagen in the tissue structure of larynx, and mainly in the vocal folds (especially in Reinke's space). This process is degenerative and especially noticeable after 60. Another fact influencing aging voice is the hormonal decay [18]. The presence of progesterone and estrogens is strongly reduced in the case of females after menopause, and testosterone decay is also noted in the case of males, although at a lesser level and extent. Another factor which introduces more uncertainty in the correlates is the consequence of alcohol, tobacco or other inhaled drug abuse affecting the upper respiratory ways. These effects worsen with age, thus influencing causing a more deteriorated phonation. These factors have to be taken into consideration to avoid misleading effects on PD grading performance.

Acknowledgments This work is being funded by grants TEC2012-38630-C04-01 and TEC2012-38630-C04-04 from Plan Nacional de I+D+i, Ministry of Economic Affairs and Competitiveness of Spain. from Plan Nacional de I+D+i, Ministry of Economic Affairs and Competitiveness of Spain. Special thanks are also due to the Patient Voice Analysis Challenge initiative for allowing the use of their data in the present study.

References

1. Dorsey, E.R., et al.: Projected number of people with Parkinson disease in the most populous nations, 2005 through 2030. Neurology **68**, 384–386 (2007)
2. Falk, T., Chan, W., Shein, F.: Characterization of atypical vocal source excitation, temporal dynamics and prosody for objective measurement of dysarthric word intelligibility. Speech Commun. **54**(5), 622–631 (2012)
3. Yunusova, Y., Weismer, G., Westbury, J., Lindstrom, M.: Articulatory movements during vowels in speakers with dysarthria. J. Speech Lang. Hear. Res. **51**, 596–611 (2008)
4. Tsanas, A.: Accurate telemonitoring of Parkinson's disease symptom severity using nonlinear speech signal processing and statistical machine leaning. Ph.D. thesis, University of Oxford, U.K., June (2012)
5. Tsanas, A., Little, M.A., McSharry, P.E., Spielman, J., Ramig, L.O.: Novel speech signal processing algorithms for high-accuracy classification of Parkinson's disease. IEEE Trans. Biomed. Eng. **59**, 1264–1271 (2010)
6. Little, M.A., McSharry, P.E., Hunter, E.J., Spielman, J., Ramig, L.O.: Suitability of dysphonia measurements for telemonitoring of Parkinson's disease. IEEE Trans. Biomed. Eng. **56**(4), 1015–1022 (2009)

7. Little, M., Wicks, P., Vaughan, T., Pentland, A.: Quantifying short-term dynamics of Parkinson's disease using self-reported symptom data from an Internet social network. J. Med. Internet Res. **15**(1), e20 (2013)
8. Hoehn, M.M., Yahr, M.D.: Parkinsonism: onset, progression, and mortality. Neurology **17**(5), 427–442 (1967)
9. Chenausky, K., MacAuslan, J., Goldhor, R.: Acoustic analysis of PD speech. Parkinson's Disease (2011). doi:10.4061/2011/435232
10. Gómez, P., et al.: Estimating tremor in vocal fold biomechanics for neurological disease characterization. In: Proceedings of the 18th International Conference on Digital Signal Processing (DSP2013), Santorini, Greece, June 2013, M1C-2
11. Gómez, P., et al.: Characterizing neurological disease from voice quality biomechanical analysis. Cogn. Comput. **5**, 399–425 (2013)
12. Schapira, A.H.V., et al.: Novel pharmacological targets for the treatment of Parkinson's disease. Nat. Rev. Drug Discov. **5**, 845–854 (2006)
13. Deller, J.R., Proakis, J.G., Hansen, J.H.L.: Discrete-Time Processing of Speech Signals. Macmillan, New York (1993)
14. Gómez, P., et al.: Glottal source biometrical signature for voice pathology detection. Speech Commun. **51**, 759–781 (2009)
15. Synapse: Contribute to the cure. https://www.synapse.org. Accessed 24 Feb 2015
16. Chang, C.C., Lin, C.J.: LIBSVM: a library for support vector machines. ACM Trans. Intell. Syst. Technol. (TIST) **2**(3), Article 27 (2011). http://www.csie.ntu.edu.tw/~cjlin/libsvm. Accessed 24 Feb 2015
17. Tsanas, A., Little, M.A., McSharry, P.E., Scanlon, B.K., Papapetropoulos, S.: Statistical analysis and mapping of the unified Parkinson's disease rating scale to Hoehn and Yahr staging. Parkinsonism Relat. Disord. **18**(5), 697–699 (2012)
18. Abitbol, J., Abitbol, P., Abitbol, B.: Sex hormones and the female voice. J. Voice **13**(3), 424–446 (1999)

Language Independent Detection Possibilities of Depression by Speech

Gábor Kiss, Miklós Gábriel Tulics, Dávid Sztahó, Anna Esposito
and Klára Vicsi

Abstract In this study, acoustic-phonetic analysis of continuous speech and statistical analyses were performed in order to find parameters in depressed speech that show significant differences compared to a healthy reference group. Read speech materials were gathered in the Hungarian and Italian languages from both healthy people and patients diagnosed with different degrees of depression. By statistical examination it was found that there are many parameters in the speech of depressed people that show significant differences compared to a healthy reference group. Moreover, most of those parameters behave similarly in other languages such as in Italian. For classification of the healthy and depressed speech, these parameters were used as an input for the classifiers. Two classification methods were compared: Support Vector Machine (SVM) and a two-layer feed-forward neural network (NN). No difference was found between the results of the two methods when trained and tested on Hungarian language (both SVM and NN classification accuracy was 75 %). In the case of training with Hungarian and testing with Italian healthy and depressed speech both classifiers reached 77 % of accuracy.

Keywords Speech analysis · Depressed speech · Pathological speech production · SVM · Neural networks

G. Kiss (✉) · M.G. Tulics · D. Sztahó · K. Vicsi
Department of Telecommunications and Media Informatics,
Budapest University of Technology and Economics, Budapest, Hungary
e-mail: kiss.gabor@tmit.bme.hu

M.G. Tulics
e-mail: sztaho@tmit.bme.hu

D. Sztahó
e-mail: tulics@tmit.bme.hu

K. Vicsi
e-mail: vicsi@tmit.bme.hu

A. Esposito
Department of Psychology and IIASS, Seconda Università di Napoli, Caserta, Italy
e-mail: iiass.annaesp@tin.it

© Springer International Publishing Switzerland 2016   103
A. Esposito et al. (eds.), *Recent Advances in Nonlinear Speech Processing*,
Smart Innovation, Systems and Technologies 48,
DOI 10.1007/978-3-319-28109-4_11

1 Introduction

Speech is a good indicator of the physiological and cognitive condition of humans, thus symptoms of depression can also be observed in speech. Experienced physicians can diagnose depression from the patient's speech quality. To name the properties of depressed speech physicians often use the words melted, slow, monotonic, lifeless and metallic. Research indicates that we can link these perceptual properties to acoustic parameters.

Many studies have determined acoustic parameters that can be connected to depression. In some of the studies, follow-up monitoring research has been carried out to gather parameters with high classification performance; others measure differences between the speech of healthy and depressed people. For the changes of mood states and emotions prosodic parameters like rhythm, intonation, accent and timing are very important [1–4].

An early study contains experiments with only three patients, but this study has already identified one of the most important acoustic parameters of depressed speech: fundamental frequency [5].

Today, researchers are studying phonetic parameters at different levels of speech production, such as: fundamental frequency, variation of fundamental frequencies, formants, power spectral density [6], cepstrum [7] or MFC coefficients (Mel-frequency cepstrum coefficients) [8], speech rate [9], glottal parameters [10], amplitude modulation and other different prosodic parameters [11–15].

Our first goal is to expand the above mentioned specific set of acoustic-phonetic parameters of depressed speech. Moreover we are going to examine the language dependency of these parameters. We look after those parameters which behave similarly in other languages such as in Italian. Our second goal is to compare two classification methods: SVM and a two-layer feed-forward NN. Furthermore we compare the classification results for the Hungarian and Italian languages.

The paper is structured as follows. The descriptions of the used databases are presented in Sect. 2, detailed descriptions of the evaluation of acoustic parameters are presented in Sect. 3, next we show our classification results in Sect. 4, followed by the conclusions in Sect. 5.

2 Databases

Read speech was used for the research, the healthy and depressed Hungarian and Italian participants were asked to read the tale "The North Wind and the Sun" in their own mother tongue. The tale is a standard phonetically balanced short folk tale (about six sentences all together), frequently used in the phoniatry practice.

2.1 Depressed Hungarian Speech Database

The Depressed Hungarian Speech Database is a collection of records from Hungarians who are suffering from depression. The database contains the speech of 54 patients (35 females and 19 males). The patients were selected with the help of a psychiatrist from the Neurology Department of Semmelweis University, Hungary. In order to measure depression and classify the recordings, Beck Depression Inventory (BDI) score was used. This is a method to specify the severity of depression in the range from 0 to 63 [16]. The categories are the following: 0–13: healthy; 14–19: mild depression; 20–28: moderate depression; 29–63: severe depression. The distribution of BDI indices of Hungarian depressed patients is shown in Fig. 1, the age distributions of the patients is shown in Fig. 2.

If a speaker's BDI score is between 0 and 13, the speaker was considered healthy, thus generating the following database.

2.2 Healthy Hungarian Speech Database

The Healthy Hungarian Speech Database contains the speech records of 73 healthy speakers (44 females and 29 males). The recordings were recorded with clip-on microphones (Audio-Technica ATR3350), with external USB sound card, with 44 100 Hz at 16 kHz sampling rate, quantized at 16 bits. The age distribution of the speakers is shown in Fig. 3.

Fig. 1 The distribution of BDI indices of the depressed people in the Hungarian database

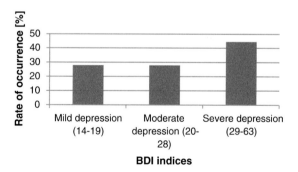

Fig. 2 The distribution of age of the depressed people in the database

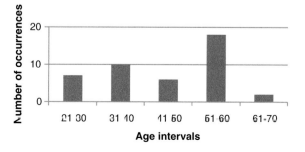

Fig. 3 Healthy reference speech database age distribution

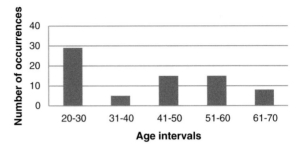

Fig. 4 The distribution of BDI indices of the depressed people in the Italian database

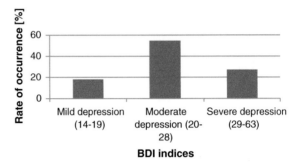

Both database recordings were annotated and segmented on phoneme level, using SAMPA phonetic alphabet with the help of an automatic phoneme segmentator developed in our Laboratory [17]. After then, manual correction was done.

2.3 Healthy and Depressed Italian Speech Database

22 recordings were collected from Italian speakers. 11 recordings are from healthy people and 11 are from patients suffering from depression. The patients who are suffering from depression were selected with the help of our Italian colleagues. The BDI indices in this case are within the range of 16 to 38. The recordings were annotated and segmented on phoneme level, the same way as the Hungarian databases. The distribution of BDI indices are shown in Fig. 4. All our databases are under continual expansion.

3 Acoustic Parameters

The acoustic-phonetic parameters were divided into two groups: segmental and supra-segmental (prosodic) parameters. The segmental parameters were calculated in the middle of the same vowel ('E'). In this article all parameters were called segmental, which were measured in the middle of the vowel, using a given time window for the calculation. These are fundamental frequency of 'E' vowels (F0), first and second formant frequency of 'E' vowels (F1, F2), jitter of the vowels 'E' (Jitter) and shimmer of the vowels 'E' (Shimmer). For the calculation of formants, fundamental frequency, jitter and shimmer, a Hamming window was used with 25 ms frame size and these parameters were always measured from the middle of each vowel 'E'. 'E' sounds were detected automatically. This vowel was chosen for the measurement, because typically is the most frequent in Hungarian and in this tale as well. There was no significant difference between the different acoustic parameters measured at different vowels (except for the vowel-specific F1 and F2 values), so it is not necessary to examine all of them, examination of the 'E' sound is sufficient.

The supra-segmental (prosodic) parameters were calculated by the total length of each recording. The following prosodic parameters were measured: intensity dynamics of speech (variance of intensity, VDS), fundamental frequency dynamics of speech (range of fundamental frequency, RFF), total length of pauses (TLP), articulation and speech rate (AR, SR), and rate of transient parts (ROT) [18].

Some of these parameters are already known as good indicators of depression: TLP, AR, SR, F1, F2, but there are others which are less or not at all used to examine the relation of depression: Jitter, Shimmer, VDS, RFF, ROT. In the examination of depression there is a very common parameter (voice onset time), which was not examined by us. This is because it is hard to calculate automatically, and our final goal is build a system which can automatically indicate depression.

For the measurement of intensity a 100 ms frame size window was used with 10 ms time step, and the variance of these values was used as VDS. For the measurement of RFF the variance of all the vowels' F0 value was used. For the measurement of the ROT 30 ms frame size window was used with 1 ms time step. The ROT value is the ratio of the total length of transient parts and the total length of the recording (except the pauses). The calculation of the transient parts is based on a specific distance method on the basis of frequency domain with Mel-band filterbank [18].

Every acoustic parameter showed normal distribution across both sexes for depressed as well as normal speech (they were tested with the Kolmogorov-Smirnov test with significance level 95 %). Two-tailed tests were used for statistical significance testing for the mean values of the acoustic parameters between depressed and healthy speech. Where F probe showed significant variances of an acoustic parameter between depressed and normal speech (with significance level 95 %), the Welch's T test was used. The significance analysis was performed separately by gender. For the T test the significant level was 90 %. The following segmental parameters showed significant differences between healthy and depressed speech at least in one gender (Table 1).

The following supra-segmental parameters showed significant differences between healthy and depressed speech (VDS, RFF, AR, SR, TLP and ROT), at least in one gender (Table 2).

Table 1 Two-sample T tests results for segmental parameters

Feature	Sex	Group	Mean	Var	P val.	Student val.	Sign. Diff.
F0 [Hz]	F	Dep.	158	26, 8	3, 5	1, 7	**YES**
		Normal	179	24, 8			
	M	Dep.	105	12, 5	1, 5	1, 7	NO
		Normal	112	16, 5			
F1 Var [Hz]	F	Dep.	68	18	0, 8	1, 7	NO
		Normal	61	13			
	M	Dep.	65	16	1, 8	1, 7	**YES**
		Normal	52	12			
F2 Var [Hz]	F	Dep.	164	28	1, 8	1, 7	**YES**
		Normal	152	31			
	M	Dep.	149	40	1, 9	1, 7	**YES**
		Normal	128	36			
Jitter [%]	F	Dep.	2, 7	0, 7	2, 6	1, 7	**YES**
		Normal	1, 3	0, 4			
	M	Dep.	3, 8	1, 3	0, 9	1, 7	NO
		Normal	3, 5	1, 1			
Jitter Var	F	Dep.	2, 3	0, 6	2, 4	1, 7	**YES**
		Normal	1, 9	0, 5			
	M	Dep.	3, 2	1	0, 7	1, 7	NO
		Normal	3	1			
Shimmer [%]	F	Dep.	11, 4	2	3, 6	1, 7	**YES**
		Normal	9, 9	1, 5			
	M	Dep.	13	1, 6	3, 1	1, 7	**YES**
		Normal	11, 5	1, 6			
Shimmer Var [%]	F	Dep.	6, 6	1, 4	3, 2	1, 7	**YES**
		Normal	5, 6	1, 1			
	M	Dep.	8, 2	1, 1	1, 4	1, 7	NO
		Normal	7, 6	1, 3			

Further investigation was carried out, how these parameters change by increasing the degree of depression. It was found that the distance from the normal value of the parameters grew according to the severity of depression. Figure 5 shows how shimmer values are affected by the severity degree of depression.

The parameters which showed significant differences between the two groups in the case of Hungarian were examined for Italian recordings as well. The two languages show similar tendencies according to theses parameters on the data obtained from the healthy and depressed populations. There was no significance analysis carried out for Italian data since our Italian databases were too small (11 healthy and 11 depressed speakers).

Table 2 Two-sample T tests results for supra-segmental parameters

Feature	Sex	Group	Mean	Var	P val.	Student val.	Sign. Diff.
VDS [dB]	F	Dep.	5, 7	1	5, 9	1, 7	**YES**
		Normal	6, 9	0, 7			
	M	Dep.	5, 6	0, 8	4, 3	1, 7	**YES**
		Normal	6, 5	0, 6			
RFF [Hz]	F	Dep.	185	55	3, 2	1, 7	**YES**
		Normal	222	42			
	M	Dep.	123	66	0, 3	1, 7	NO
		Normal	129	48			
AR [count/s]	F	Dep.	12	1, 5	0, 9	1, 7	NO
		Normal	12, 5	1, 2			
	M	Dep.	12	2	3, 4	1, 7	**YES**
		Normal	14	1			
SR [count/s]	F	Dep.	10, 4	1, 5	0, 4	1, 7	NO
		Normal	10, 9	1, 3			
	M	Dep.	10	3	3, 7	1, 7	**YES**
		Normal	11, 7	2, 7			
TLP [s]	F	Dep.	8, 1	0, 6	0, 2	1, 7	NO
		Normal	7, 9	0, 5			
	M	Dep.	10, 7	2	3, 2	1, 7	**YES**
		Normal	7, 1	0, 9			
ROT [%]	F	Dep.	23	2, 1	1, 8	1, 7	**YES**
		Normal	26	2, 3			
	M	Dep.	20	2, 9	3	1, 7	**YES**
		Normal	26	2, 2			

Fig. 5 Shimmer distribution in function of BDI values in the case of male speakers

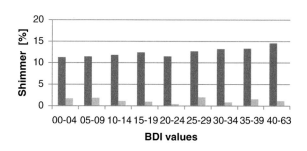

4 Classifications

Our main goal was to detect depression by speech, and we intended to build a system which could work mostly in a language-independent way. For this reason, first, acoustic parameters were selected according to our statistical examination in Sect. 3, which showed significant difference at least in one gender. Fundamental frequency values were not used because they strongly vary from one person to another. Thereafter, parameters were selected which, on the base of our phonetic knowledge, operate language independently. The first and second formant frequencies, articulation and speech rate are surely language-dependent parameters, these parameters were left out as well. Thus, the following parameters were used for the classification tasks: range of fundamental frequency (RFF), volume dynamics of speech (VDS), the variance of the first and the second formant frequency (F1 Var, F2 Var), jitter (Jitter), shimmer (Shimmer), the variance of jitter and shimmer (Jitter Var, Shimmer Var), rate of transient (ROT), and the total length of pauses (TLP), the variance of the first and the second formant frequency (F1 Var, F2 Var), jitter (Jitter), shimmer (Shimmer), the variance of jitter and shimmer (Jitter Var, Shimmer Var) were calculated at the middle of vowels 'E' in case of Hungarian, as described in Session 3. In the case of Italian these segmental parameters were measured instead that from the "E" from the vowel 'I', because this vowel was the most frequent one in the Italian fairy tale. We could rightly do this, because as we have mentioned it in the Chap. 4, we did not find significant differences among the selected parameters in the case of different vowels. The mean and/or variance values of each parameter were calculated for each recording and were used for the classification.

Two classification methods were compared first for Hungarian and then for Italian databases: SVM (using kernel type of radial basis function) and NN. To perform classifications with SVM, LIBSVM integrated software version 3.20 was used [19]. For NN type classifications a two-layer feed-forward network was created (with sigmoid hidden and softmax output neurons) with the help of the Matlab Neural Network Toolbox [20].

The two classifiers (SVM, NN) were trained and tested with the Hungarian databases. A total of 108 recordings were used for the classification task, 54 recordings from healthy and 54 recordings from depressed readers. The database was distributed in three parts: 55 % of the data was used for training, 15 % for development (to optimize parameters) and 30 % for testing. The training set contained 30 recordings from healthy and 30 recordings from depressed readers, the development set contained 8 recordings from healthy and 8 recordings from depressed readers and the test set contained 16 recordings from healthy and 16 recordings from depressed readers.

With the balanced train, development and test set, a random baseline classifier would give us 50 %. With the experiments on the development set, the best classifier parameters were selected (parameters where the classification results on the development set were close to the best results) and these parameters were used to determine the classification accuracy on the test set. Our best classification results are shown in Table 3.

Table 3 Classification results using SVM and neural network (Training data: Hungarian; Testing data: Hungarian)

Classifier	Acoustic parameters	C	Gamma	Number of hidden nodes	Classification accuracy (%)
SVM	F1 Var, F2 Var, Jitter, Jitter Var, Shimmer, Shimmer Var, RFF, VDS, TLP, ROT	4	0, 125	–	75
NN	F1 Var, F2 Var, Jitter, Jitter Var, Shimmer, Shimmer Var, RFF, VDS, TLP, ROT	–	–	18	75

Table 4 Classification results using SVM and neural network (Training data: Hungarian; Testing data: Italian)

Classifier	Acoustic parameters	C	Gamma	Number of hidden nodes	Classifiction accuracy (%)
SVM	F1 Var, F2 Var, Jitter, Jitter Var, Shimmer, Shimmer Var, RFF, VDS, TLP, ROT	4	0, 125	–	77
NN	F1 Var, F2 Var, Jitter, Jitter Var, Shimmer, Shimmer Var, RFF, VDS, TLP, ROT	–	–	18	76

No difference was found between the result of the two methods when trained and tested on Hungarian. In both cases the best accuracy was 75 %. When searching for the best result, the numbers of hidden neurons were mapped between 2 to twice the number of inputs, in our case 20 hidden neurons. The best result was obtained at 18 hidden neurons when training and testing on Hungarian. We wanted to examine whether the selected parameters could be used in other languages too, for the detection of depression. Thus, the classifiers were trained with the Healthy Hungarian Speech and Depressed Hungarian Speech databases (with the same training and parameter combination used for the first classification). Testing was done with the Italian recordings, presented in Sect. 2.3. The classification results are shown in Table 4.

It was found that SVM and NN gave slightly better results when trained and tested on the Italian dataset. Both classifiers categorized 17 correct from the 22 total recordings. The increased classification results may be misleading due to differences in the test sets, although correct or incorrect classification of a record causes high percentage deviation due to the limited number of samples.

Note that the BDI scores of the Hungarian and Italian depressed databases differ and the number of samples limits the significance of the results in both databases. However, it is interesting to note that the classification results, when testing with Italian dataset, did not change significantly, despite the fact that the classifier was trained by Hungarian.

5 Conclusions

In this study we performed statistical examination in order to identify acoustic-phonetic parameters associated with depression that show significant differences compared to a healthy reference group. It was found that there are many parameters in the speech of depressed people that show significant differences compared to a healthy reference group. Our selected parameters are partly different from those used in previous works. Our 75 and 77 % of correct classification percentages were slightly better than the 72 % achieved by David et al. (2013, [21]), even though results are not comparable since the databases were different.

The selected parameters were examined on a small Italian database too, which contained both healthy and depressed speech. It was found that the differences of the selected parameters' values, between the healthy and depressed speech, were similar in case of Hungarian and Italian language. On the basis of the results of the classification experiment with the Hungarian and Italian databases, it can be said that those 10 selected speech parameters significantly change in the case of depression, and moreover, these parameters behave similarly in the case of Italian language too and possibly in the case of other languages. Of course, the proof of this statement requires that the investigation should be extended to other languages as well. These parameters are as follows: range of fundamental frequency, volume dynamics of speech, the variance of the first and the second formant frequency, jitter, shimmer, the variance of jitter and shimmer, rate of transient, and the total length of pauses.

In addition, we compared two classification methods: SVM and a two-layer feed-forward NN on our databases. The selected parameters, which indicated depression in the statistical examination, were not language dependent when used as an input for the classifiers for classification of the healthy and depressed speech. No difference was found between the result of the two methods when trained and tested on Hungarian; we gained 75 % accuracy with both classifiers. When the classifiers were trained by Hungarian and tested with Italian dataset SVM and NN gave slightly better results. The classification accuracy was 77 % in both cases. We can presume that the separation of healthy and depressed speech carries similar tendencies for both Hungarian and Italian languages.

The presented results clearly indicate that the development of such a medical device able to diagnose and identify depressive states is quite realistic and worth working for it. Of course we need much more data and the method needs to be tested by further languages.

Acknowledgments The authors would like to thank European Space Agency COALA project: Psychological Status Monitoring by Computerised Analysis of Language phenomena (COALA) (AO-11-Concordia).

References

1. Ivry, R.B., Justus, T.C., Middleton, C.: The cerebellum, timing, and language: implications for the study of dyslexia. In: Wolf, M. (ed.), Dyslexia Fluency and the Brain, pp. 198–211. York Press, Timonium, MD (2001)
2. Esposito, A., Bourbakis, N.: The role of timing in speech perception and speech production processes and its effects on language impaired individuals. In: Sixth IEEE Symposium on BioIn-formatics and BioEngineering, 2006. BIBE 2006. IEEE (2006)
3. Vicsi, K., Sztahó, D.: Problems of the automatic emotion recognitions in spontaneous speech; an example for the recognition in a dispatcher center. In: Esposito, A., Martone, R., Müller, V., Scarpetta, G. (eds.) Toward Autonomous, Adaptive, and Context-Aware Multimodal Interfaces. Theoretical and Practical Issues, vol. 6456, pp. 331–339. Springer, Heidelberg (2011)
4. Tóth, S.Z.L., Sztahó, D., Vicsi, K.: Speech emotion perception by human and machine. In: Proceeding of COST Action 2102 International Conference: Revised Papers in Verbal and Non-verbal Parameters of Human-Human and Human-Machine Interaction, pp. 213–224. Springer, Patras (2007)
5. Askenfelt, A., Sjoelin, A.: Voice analysis in depressed patients: rate of change of fundamental frequency related to mental state. Speech Transmission Laboratory—Quarterly Progress and Status Report, pp. 71–84. Royal Institute of Technology, Stockholm (1980)
6. Daniel, J., at all: Acoustical properties of speech as indicators of depression and suicidal risk. In: IEEE Transactions on Biomedical Engineering, vol. 47, no. 7 (2000)
7. Thaweesak, Y., et al.: Characterizing sub-band spectral entropy based acoustics as assessment of vocal correlate of depression. In: International Conference on Control, Automation and Systems, 27–30 Oct 2010
8. Terapong, B., et al.: Assessment of vocal correlates of clinical depression in female subjects with probabilistic mixture modeling of speech cepstrum. In: 2011 11th International Conference on Control, Automation and Systems, 26–29 Oct 2011
9. James, C., et al.: Voice acoustic measures of depression severity and treatment response collected via interactive voice response (IVR) technology. J. Neurolinguistics (2007)
10. Elliot, M., et al.: Investigating the role of glottal parameters in classifying clinical depression. In: Proceedings of the 25th Annual International Conference of the IEEE, pp. 2849–2852 (2003)
11. Sanchez, M.H., et al.: Using prosodic and spectral parameters in detecting depression in elderly males. INTERSPEECH 2011. Florence, Italy, 27–31 Aug 2011
12. Alghowinem, S., Goecke, R., Wagner, M., Epps, J., Breakspear, M., Parker, G.: Detecting depression—a comparison between spontaneous and read speech. In: 38th International Conference on Acoustics, Speech, and Signal Processing (ICASSP) (2013)
13. Helfer, B.S., Quatieri, T.F., Williamson, J.R., Mehta, D.D., Horwitz, R., Yu, B.: Classification of depression state based on articulatory precision. In: 14th Annual Conference of the International Speech Communication Association (2013)
14. Mundt, J.C., Snyder, P.J., Cannizzaro, M.S., Chappie, K., Geralts, D.S.: Voice acoustic measures of depression severity and treatment response collected via interactive voice response (IVR) technology. J. Neurolinguistics (2007)
15. Kiss, G., Vicsi, K.: Physiological and cognitive status monitoring on the base of acoustic-phonetic speech parameters. In: Besacier, L., Dediu, A.-H., Martín-Vide, C. (eds.) Lecture Notes in Computer Science: Statistical Language and Speech Processing. Grenoble, France, 14–16 Oct 2014

16. Abela, J.R.Z., D'Allesandro, D.U.: Beck's cognitive theory of depression: the diathesis-stress and causal mediation components. Br. J. Clin. Psychol. **41**, 111–128 (2002)
17. Kiss, G., Sztahó, D., Vicsi, K.: Language independent automatic speech segmentation into phoneme-like units on the base of acoustic distinctive features. In: 4th IEEE International Conference on Cognitive Infocommunications—CogInfoCom 2013. Budapest, Hungary, 2013.12.02–2013.12.06
18. Kiss, G., Sztahó, D., Vicsi, K., Golemis, A.: Connection between body condition and speech parameters—especially in the case of hypoxia. In: 5th IEEE International Conference on Cognitive Infocommunications (CogInfoCom 2014), pp. 333–336. Vietri, Italy, 05–07 Nov 2014
19. Chang, C.-C., Lin, C.-J.: LIBSVM: a library for support vector machines. ACM Trans. Intell. Syst. Technol. 2:27:1–27:27 (2011). Software available at http://www.csie.ntu.edu.tw/~cjlin/libsvm
20. Beale, M.H., Hagan, M.T., Demuth, H.B.: Neural Network Toolbox, User's Guide. The Mathworks Inc. (2010)
21. DeVault, D., Georgila, K., Artstein, R., Morbini, F., Traum, D., Scherer, S., Skip Rizzo, A., Morency, L.-P.: Verbal indicators of psychological distress in interactive dialogue with a virtual human. In: The 14th Annual Meeting of the Special Interest Group on Discourse and Dialogue (SigDial 2013), pp. 193–202. Metz, France, Aug 2013

Part IV
Improving VUI

Constructing a Deep Neural Network Based Spectral Model for Statistical Speech Synthesis

Shinji Takaki and Junichi Yamagishi

Abstract This paper presents a technique for spectral modeling using a deep neural network (DNN) for statistical parametric speech synthesis. In statistical parametric speech synthesis systems, spectrum is generally represented by low-dimensional spectral envelope parameters such as cepstrum and LSP, and the parameters are statistically modeled using hidden Markov models (HMMs) or DNNs. In this paper, we propose a statistical parametric speech synthesis system that models high-dimensional spectral amplitudes directly using the DNN framework to improve modelling of spectral fine structures. We combine two DNNs, i.e. one for data-driven feature extraction from the spectral amplitudes pre-trained using an auto-encoder and another for acoustic modeling into a large network and optimize the networks together to construct a single DNN that directly synthesizes spectral amplitude information from linguistic features. Experimental results show that the proposed technique increases the quality of synthetic speech.

1 Introduction

Recently, deep neural networks (DNNs) with many hidden layers have been significantly improved in statistical speech synthesis researches. For instance, DNNs have been applied for acoustic modelling. Zen et al. [1] use DNN to learn the relationship between input texts and extracted features instead of decision tree-based

---

Shinji Takaki was supported in part by NAVER Labs.
Junichi Yamagishi—The research leading to these results was partly funded by EP/J002526/1 (CAF).

S. Takaki (✉) · J. Yamagishi
National Institute of Informatics, Tokyo, Japan
e-mail: takaki@nii.ac.jp

J. Yamagishi
The Centre for Speech Technology Research, University of Edinburgh, Edinburgh, UK
e-mail: jyamagis@nii.ac.jp

© Springer International Publishing Switzerland 2016
A. Esposito et al. (eds.), *Recent Advances in Nonlinear Speech Processing*,
Smart Innovation, Systems and Technologies 48,
DOI 10.1007/978-3-319-28109-4_12

state tying. Restricted Boltzmann machines or deep belief networks have been used to model output probabilities of hidden Markov model (HMM) states instead of GMMs [2]. Recurrent neural network and long-short term memory have been used for prosody modelling [3] and acoustic trajectory modelling [4]. In addition, an auto-encoder neural network has also been used to extract low dimensional excitation parameters [5].

However, the synthetic speech of the latest statistical parametric speech synthesis still sounds muffled, and averaging effects of statistical models are often said to remove spectral fine structure of natural speech. To improve the quality of synthetic speech, a stochastic postfilter approach has been proposed [6] where a DNN is used to model the conditional probability of the spectral differences between natural and synthetic speech. The approach was found to be able to reconstruct the spectral fine structure lost during modeling and has significantly improved the quality for synthetic speech [6]. In this experiment, the acoustic model was trained using lower dimensional spectral envelope features, while the DNN-based postfiler was trained using the spectral amplitudes obtained using the STRAIGHT vocoder [7]. From the experimental findings, we can hypothesize that the current statistical parametric speech synthesis may suffer from quality loss due to not only statistical averaging but also acoustic modeling using lower dimensional acoustic features.

On the basis of this hypothesis, in this paper we present a new technique for constructing a DNN that directly synthesizes spectral amplitudes from linguistic features without using spectral envelope parameters such as mel-cepstrum. It is well known that there are many problems for training a DNN such as the local optima, vanishing gradients and so on [8]. However, it has been reported in the ASR field that DNNs that deal with high-dimensional features, e.g. FFT frequency spectrum, can be appropriately constructed using an efficient training technique such as pre-training [9].

Thus, in this paper we propose an efficient training technique for constructing a DNN that directly synthesizes spectral amplitudes from input texts. A key idea is to stack two DNNs, an auto-encoder neural network for data-driven nonlinear feature extraction from the spectral amplitudes and another network for acoustic modeling and context clustering. The proposed technique is regarded as a function-wise pre-training technique for constructing the DNN-based speech synthesis system.

The rest of this paper is organized as follows. Section 2 reviews a DNN-based acoustic model for the statistical parametric speech synthesis. Section 3 describes a DNN-based acoustic feature extractor and spectrum re-generator. Section 4 explains the proposed technique for constructing a DNN that directly synthesizes the spectral amplitudes. The experimental conditions and results are shown in Sect. 5. Concluding remarks and future works are presented in Sect. 6.

Fig. 1 A framework of DNN-based acoustic model

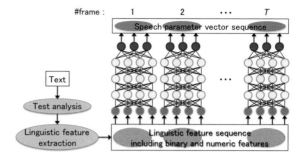

2 DNN-based Acoustic Model for Statistical Parametric Speech Synthesis

It is believed that the human speech production system has layered hierarchical structures to convert the linguistic information into speech. To approximate such a complicated process, DNN-based acoustic models that represent the relationship between linguistic and speech features have been proposed for statistical parametric speech synthesis [1–4] This section briefly reviews one of the state-of-the-art DNN-based acoustic models [1].

Figure 1 illustrates a framework of the DNN-based acoustic model. In this framework, linguistic features obtained from a given text are mapped to speech parameters by a DNN. The input linguistic features include binary answers to questions about linguistic contexts and numeric values, e.g. the number of words in the current phrase, the position of the current syllable in the word, and durations of the current phoneme. In [1], the output speech parameters include spectral and excitation parameters and their time derivatives (dynamic features). By using pairs of input and output features obtained from training data, the parameters of the DNN can be trained with a stochastic gradient descend (SGD) [10]. Speech parameters can be predicted for an arbitrary text by a trained DNN using forward propagation.

3 Deep Auto-encoder Based Acoustic Feature Extraction

An auto-encoder is an artificial neural network that is used generally for learning a compressed and distributed representation of a dataset. It consists of the encoder and the decoder. In the basic one-hidden-layer auto-encoder, the encoder maps an input vector $\mathbf{x}$ to a hidden representation $\mathbf{y}$ as follows:

$$\mathbf{y} = f_\theta(\mathbf{x}) = s(\mathbf{W}\mathbf{x} + \mathbf{b}), \tag{1}$$

where $\theta = \{\mathbf{W}, \mathbf{b}\}$. $\mathbf{W}$ and $\mathbf{b}$ represent an $m \times n$ weight matrix and a bias vector of dimensionality m, respectively, where n is the dimension of $\mathbf{x}$. The function s is a non-linear transformation on the linear mapping $\mathbf{Wx} + \mathbf{b}$. A sigmoid, a tanh, or a relu function is typically used for s. $\mathbf{y}$, the output of the encoder, is then mapped to $\mathbf{z}$, the output of the decoder. The mapping is performed by a linear mapping followed by an arbitrary function t that employs an $n \times m$ weight matrix $\mathbf{W}'$ and a bias vector of dimensionality n as follows:

$$\mathbf{z} = g_{\theta'}(\mathbf{y}) = t(\mathbf{W}'\mathbf{y} + \mathbf{b}'), \tag{2}$$

where $\theta' = \{\mathbf{W}', \mathbf{b}'\}$. An auto-encoder can be made deeper by stacking multiple layers of encoders and decoders to form a deep architecture.

Pre-training is widely used for constructing a deep auto-encoder. In pre-training, the number of layers in a deep auto-encoder increases twice as compare to a deep neural network (DNN) when stacking each pre-trained unit. It has been reported that fine-tuning with back-propagaqion through a deep auto-encoder is ineffective due to vanishing gradients at the lower layers [8]. To overcome this issue, we restrict the decoding weight as the transpose of the encoding weight following [10], that is, $\mathbf{W}' = \mathbf{W}^T$ where $\mathbf{W}^T$ denotes the transpose of $\mathbf{W}$. Each layer of a deep auto-encoder can be pre-trained greedily to minimize the reconstruction loss of the data locally. Figure 2 shows a procedure for constructing a deep auto-encoder using pre-training. In pre-training, a one-hidden-layer auto-encoder is trained and the encoding output of the locally trained layer is used as the input for the next layer. After all layers are pre-trained, they are stacked and are fine-tuned to minimize the reconstruction error over the entire dataset using error backpropagation [11]. We use the mean square error (MSE) for the loss function of a deep auto-encoder.

Figure 3 shows an example of original and reconstructed spectrograms using the standard mel-cepstral analysis and a deep auto-encoder. Both mel-cepstral analysis and the deep auto-encoder produced 120-dimensional acoustic features. We can clearly see that the deep auto-encoder reconstructs spectral fine structures more precisely than that of the mel-cepstral analysis. Log spectral distortions between natural spectrum and reconstructed spectrum calculated using 441 sentences were

Fig. 2 Greedy layer-wise pre-training for constructing a deep auto-encoder

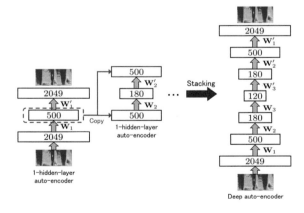

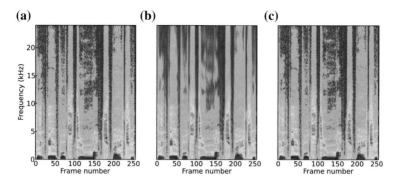

Fig. 3 Original and reconstructed spectra using mel-cepstral analysis and a deep auto-encoder. **a** Original, **b** mel-cepstrum, **c** deep auto-encoder

2.53 and 1.19 dB for the mel-cepstral analysis and deep auto-encoder, respectively. Similar auto-encoder based bottleneck features were tested for a ClusterGen speech synthesizer [12]. Our idea is different from [12] and we stack the decoder part of the deep auto-encoder onto another DNN for acoustic modeling.

4 Proposed DNN-based Spectral Modeling

A DNN-based acoustic model described in Sect. 2 may be used for the direct spectral modeling by substituting an output of the network from mel-cepstrum to the spectrum. However, the dimension of spectrum is much higher than that of mel-cepstrum. For a speech signal at 48 kHz, the mel-cepstral analysis order typically used is around 50-dim, whereas the dimension of spectrum corresponds to FFT points such as 2049. Because of this high dimensional data, a more efficient training technique is needed to construct a DNN that directly represents the relationship between linguistic features and spectra. In this paper, we hence propose a function-wise pre-training technique where we explicitly divide the general flow of the statistical parametric speech synthesis system into a few sub-processes, construct and optimize a DNN for each task individually, and stack the individual networks for the final optimization.

Figure 4 shows a procedure for constructing the proposed DNN-based spectral model. Details of each step of the proposed technique are as follows:

Step 1. Train a deep auto-encoder using spectra and extract bottleneck features for a DNN-based acoustic model used in Step 2. Layer-wise pre-training or other initialization may be used for the learning of the deep auto-encoder.
Step 2. Train a DNN-based acoustic model using the bottleneck features extracted in Step 1. Layer-wise pre-training or other initialization may be used for learning the DNN.

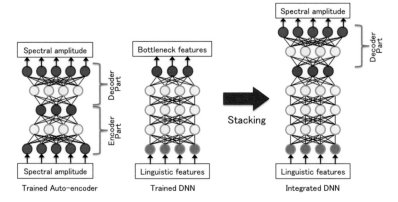

Fig. 4 Constructing a DNN-based spectral model based on a deep autoencoder and a DNN-based acoustic model

Step 3. Stack the trained DNN-based acoustic model for bottleneck features and the decoder part of the trained deep auto-encoder as shown in Fig. 4 and optimize the whole network.

A DNN that represents the relationship between linguistic features and spectra is constructed based on a DNN-based spectral generator and a DNN-based acoustic model using the bottleneck features. After this proposed pre-training, we fine-tune the DNN to minimize the error over the entire dataset using pairs of linguistic features and spectra in training data with SGD.

5 Experiments

We have evaluated the proposed technique in the subjective experiment. The dataset we use consists of 4546 short audio waveforms uttered by a professional female native speaker of English and each waveform is around 5 s long. All data was sampled at 48 kHz.

We compared three techniques; *CEPSTRUM* is the DNN that synthesizes cepstrum vectors, *SPECTRUM* has the same network structure as that of *CEPSTRUM*, but it outputs the spectral amplitudes directly, and *INTEG* is the proposed DNN that synthesizes spectrum amplitudes with the proposed pre-training framework. In these techniques, the dynamic and acceleration features were not used. Figure 5 shows structures of constructed DNNs for each technique. We trained five-hidden-layer DNN-based acoustic models for each technique. The number of units in each of the hidden layers was set to 1024. Random initialization was used in a way similar to [1]. In *INTEG*, we trained the symmetric five-hidden-layer auto-encoder. The numbers of units of the hidden layers were 2049, 500, 60, 500 and 2049 As a result, we

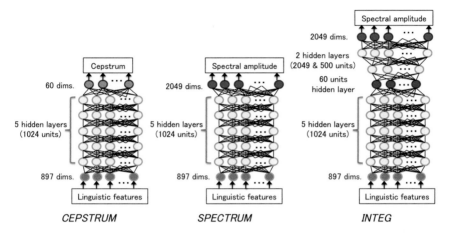

Fig. 5 Structures of constructed DNNs for each technique

constructed and fine-tuned the eight-hidden-layer (1024-1024-1024-1024-1024-60-500-2049) DNN for *INTEG*. We used a sigmoid function for all units of hidden and output layers of all DNNs.

For each waveform, we first extract its frequency spectra using the STRAIGHT vocoder with 2049 FFT points. For constructing the conventional system, 59 dimensional cepstrum coefficients were used. Spectrum and cepstrum were both frequency-warped using the Bark scale. Note that all the techniques synthesize only spectrum features and other requisite acoustic features; that is, F0 and aperiodicity measures were synthesized from the same HMM-based synthesis system [13]. Feature vectors for HMMs were comprised of 258 dimensions: 59 dimensional bark-cepstral coefficients (plus the 0th coefficient), log f0, 25 dimensional band aperiodicity measures, and their dynamic and acceleration coefficients. Phoneme durations were also estimated by HMM-based speech synthesis. The context-dependent labels were built using the pronunciation lexicon Combilex [14]. The linguistic features for DNN acoustic models were comprised of 897 dimensions: 858 dimensional binary features for categorical linguistic contexts, 36 numerical features for numerical linguistic contexts, and three numerical features for the position of the current frame and duration of the current segment. The linguistic features and spectral amplitudes in the training data were normalized for training DNNs. In the proposed technique, however, the bottleneck features are not normalized, and the normalization process is not used for hidden units in the integrated DNN. The input linguistic features were normalized to have zero-mean unit-variance, whereas the output spectral amplitudes were normalized to be within 0.0–1.0.

We synthesized speech samples from spectrum amplitudes, F0 features and aperiodicity measures using the STRAIGHT vocoder in all techniques. In *CEPSTRUM*, synthesized cepstral vectors were converted into spectrum amplitudes for using the STRAIGHT vocoder.

Fig. 6 Results of preference test

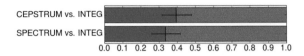

In subjective experiments, two preference tests were conducted. Seven subjects participated in both listening tests. Thirty sentences were randomly selected from the 180 sentences for each subject. The experiment was carried out using headphones in a quiet room.

5.1 Experimental Result

Figure 6 shows the results of the preference tests with 95% confidence intervals. In the first preference test, they were asked to compare the DNN that synthesizes cepstrum vectors (*CEPSTRUM*) with the proposed DNN (*INTEG*). In the second preference test, they were asked to compare the DNN without the proposed pre-training technique that synthesizes spectrum amplitudes (*SPECTRUM*) with the proposed DNN (*INTEG*). The figure shows that the proposed technique produces more natural-sounding speech than other techniques. This indicates that the DNN that directly synthesizes spectra was efficiently trained using the proposed technique.

6 Conclusion

In this paper, we have proposed a technique for constructing a DNN that directly synthesizes spectral amplitudes. On the basis of the general flow for constructing the statistical parametric speech synthesis systems, a part of layers of a DNN could be efficiently pre-trained. Experimental results showed that the proposed technique increased the quality of synthetic speech.

In future work, we will investigate the effect of structures of a DNN-based acoustic model and a DNN-based spectrum auto-encoder more thoroughly. Time derivative features will also be interesting to investigate.

References

1. Zen, H., Senior, A., Schuster, M.: Statistical parametric speech synthesis using deep neural networks. In: Proceedings of ICASSP, pp. 7962–7966 (2013)
2. Ling, Z.-H., Deng, L., Yu, D.: Modeling spectral envelopes using restricted Boltzmann machines and deep belief networks for statistical parametric speech synthesis. IEEE Trans. Audio Speech Lang. Process. **21**, 2129–2139 (2013)

3. Fan, Y., Qian, Y., Xie, F., Soong, F.K.: TTS synthesis with bidirectional LSTM based recurrent neural networks. In: Proceedings of Interspeech, pp. 1964–1968 (2014)
4. Fernandez, R., Rendel, A., Ramabhadran, B., Hoory, R.: Prosody contour prediction with long short-term memory, bi-directional, deep recurrent neural networks. In: Proceedings of Interspeech, pp. 2268–2272 (2014)
5. Vishnubhotla, R., Fernandez, S., Ramabhadran, B.: An autoencoder neural-network based low-dimensionality approach to excitation modeling for hmm-based text-to-speech. In: Proceedings of ICASSP, pp. 4614–4617 (2010)
6. Chen, L.-H., Raitio, T., Valentini-Botinhao, C., Yamagishi, J., Ling, Z.-H.: DNN-based stochastic postfilter for HMM-based speech synthesis. In: Proceedings of Interspeech, pp. 1954–1958 (2014)
7. Kawahara, H., Masuda-Katsuse, I., Cheveigne, A.: Restructuring speech representations using a pitch-adaptive time-frequency smoothing and an instantaneous-frequency-based F0 extraction: Possible role of a repetitive structure in sounds. Speech Commun. 27, 187–207 (1999)
8. Hochreiter, S., Bengio, Y., Frasconi, P., Schmidhuber, J.: Gradient flow in recurrent nets: the difficulty of learning long-term dependencies. Citeseer (2001)
9. Hinton, G.E.: Learning multiple layers of representation. Trends Cogn. Sci. 11, 428–434 (2007)
10. Hinton, G.E., Salakhutdinov, R.: Reducing the dimensionality of data with neural networks. Science 313(5786), 504–507 (2006)
11. Rumelhart, D.E., Hinton, G.E., Williams, R.J.: Parallel Distributed Processing: Explorations in the Microstructure of Cognition, vol. 1, pp. 318–362 (1986)
12. Muthukumar, P.K., Black, A.: A deep learning approach to data-driven parameterizations for statistical parametric speech synthesis (2014). arXiv:1409.8558
13. Zen, H., Tokuda, K., Black, A.W.: Statistical parametric speech synthesis. Speech Commun. 51, 1039–1064 (2009)
14. Richmond, K., Clark, R., Fitt, S.: On generating combilex pronunciations via morphological analysis. In: Proceedings of Interspeech, pp. 1974–1977 (2010)

The Influence of Adaptation Database Size on the Quality of HMM-based Synthetic Voice Based on the Large Average Voice Model

Martin Sulír and Jozef Juhár

Abstract This paper describes the influence of various database size on the overall quality of HMM-based synthetic voices which were built with the help of large average voice model using an average voice-based speech synthesis system together with AHO-coder vocoding technique. Together, eight new voices of one male speaker were built by gradually adding new data into adaptation database while the quality of individual new voices were evaluated with the help of objective evaluation methods. A mean mel-cepstral distortion together with an aligned and averaged fundamental frequency assessment were used for the evaluation of newly created voices. The aim was to show the effect of adding more data into adaptation database to the overall quality as well as to define the threshold when the impact of the newly added data on the voice quality will be negligible in case of using a sufficiently large average voice model in the adaptation procedure. The result of this work is a set of spectrum and fundamental frequency assessments which directly show the dependence of the voice quality on the amount of adaptation data together with the threshold definition.

Keywords Hidden Markov models · Models adaptation · Objective evaluation · Statistical parametric speech synthesis · Text-to-speech

1 Introduction

A Speech synthesis, which is represented by the Text-To-Speech (TTS) systems, is one of the most important part of the speech interaction with the computer. The main task of these systems is primarily making life easier for example to people with physical disabilities, such as blind people, or to totally ordinary people and

M. Sulír (✉) · J. Juhár
Department of Electronics and Multimedia Communications,
Technical University of Košice, Letná 9, 042 00 Košice, Slovakia
e-mail: martin.sulir@tuke.sk

J. Juhár
e-mail: jozef.juhar@tuke.sk

© Springer International Publishing Switzerland 2016   127
A. Esposito et al. (eds.), *Recent Advances in Nonlinear Speech Processing*,
Smart Innovation, Systems and Technologies 48,
DOI 10.1007/978-3-319-28109-4_13

facilitating their day to day operations [1]. The expansion of such systems was mainly caused by the demand for speech as one of the modalities, especially in the interactive applications where the speech communication with the device facilitates the transfer of information from system to user. Nowadays, the hidden Markov model based speech synthesis method represents one of the most progressive approach how to convert written text into sound, which ultimately sounds like human speech [2]. The progressiveness of this method is particularly involved in its high flexibility, where it allows to quite easily create the new voices with the help of adaptation, interpolation or, for example, using the technique of eigenvoice. The utilization of these techniques arise from using of hidden Markov models (HMM) which can be properly mathematically modified in order to obtain their desired modified versions. Models adaptation provides a relatively large range for customizing of models, where it uses a principle of using a small adaptation database and the large pre-trained average voice model to create a new voice which corresponds to the speaker who recorded the input adaptation database. In this case, the adapted voice quality is influenced by the two major factors: the adaptation database size and its quality and the average voice model diversity. The main motivation for the experiments described in this paper was to verify the dependence between the large diverse average voice model and various sized adaptation databases as well as to determine their impact on the overall adapted voice quality at the output of the system. The paper is organized as follows: in Sect. 2 an average voice-based speech synthesis (AVSS) system is described. Section 3 describes a speech synthesis system for evaluation. In Sect. 4, the experiments and results of objective evaluation are presented. The conclusions are listed at the end of this paper.

2   Average Voice-based Speech Synthesis System

The adaptation techniques were firstly developed for the automatic speech recognition systems, where they enable to adapt general acoustic models to a specific topic, speaker or environment which increases the accuracy of a speech recognition. The basic principles of these techniques have been applied also into speech synthesis systems where they showed their potential in a speaker adaptation task. The average voice-based speech synthesis system builds upon the original HMM-based system, which has been extended by a process of model adaptation [3]. Figure 1 shows a block diagram of the AVSS system for the HMM-based speech synthesis with a speaker adaptation.

The AVSS system can be divided into three basic parts. The first one is a training part which consists of the average models training. Its main task is an extraction of spectral and excitation parameters from the speech database as well as an implementation of the HMMs training. In the case of the average models training, the speech database consists of the multiple sub-databases, where each of them was recorded by one certain speaker. In the HMM-based speech synthesis method, each of the HMMs correspond to a left-to-right model where each output vector is composed of

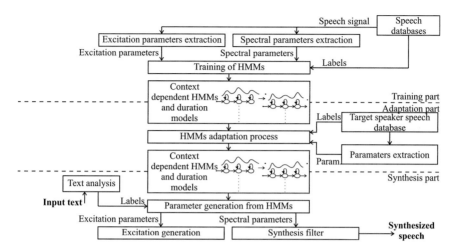

Fig. 1 Block diagram of AVSS system for HMM-based speech synthesis

two components, namely it consists of the spectrum part, represented by mel-cepstral coefficients and their delta and delta-delta coefficients and the excitation part which is represented by the excitation parameters and their corresponding delta and delta-delta dynamic features. The trained HMMs include also the density distributions for state duration for the purposes of reproduction of the temporal structure of speech. A conventional HMM-based TTS system works as a mel-cepstral vocoder with a simple impulse train as the excitation signal, where a sequence of periodic pulses and white noise together with a Mel log spectrum approximation (MLSA) filter are used. The speech synthesis process is carried out by filtering of the pulse train in case of the voiced segments and white noise filtering in case of the unvoiced speech segments where the excitation is controlled by the logarithmic values of the generated fundamental frequency. The filter parameters are adjusted according to the input mel-cepstral coefficients obtained in the parameter generation process. A several high-quality vocoders with a more advanced excitation were implemented into HMM-based speech synthesis system. Such methods include, e.g., MELP (Mixed Excitation Linear Prediction) method [4], HSM (Harmonic/Stochastic Model) model [5], excitation model based on modeling of residues [6], STRAIGHT (Speech Transformation and Representation using Adaptive Interpolation of Weighted Spectrum) [7] or AHOcoder [8]. The latter approach is one of the most advanced vocoder therefore the experiments in this work were carried out with this vocoding method. The second part of AVSS system is an adaptation step. A target speaker speech database together with its spectral and excitation parameters represent inputs to the adaptation process. The process of adaptation of the average voice models is implemented in this part, where a lot of modifications of the basic adaptation methods (Maximum likelihood linear regression, etc.) have been developed [9]. In general, the multiple linear transformation functions are estimated with the help of target speaker data during the adaptation process. However, it is not possible to estimate a transformation function for each context dependent model so each function is usually shared by a group of related models. A synthesis part of the average voice HMM-based speech synthesis system consists of two main components. The first component is represented by the text analyzer, which convert a given text into contextual label sequence. The second component consists of the several blocks which are responsible for the parameter generation from the context dependent HMMs and duration models; excitation generation based on the generated excitation parameters and the synthesis filter. This component compose a HMM sequence by concatenating context-dependent HMMs according to input label sequence. Subsequently, the state durations for the concatenate HMM sequence are determined in order to maximize the output probability of the state durations. The vectors of mel-cepstral coefficients and logarithmic values of generated fundamental frequency values are generated based on the obtained HMM sequence and the speech waveform is synthesized from these vectors by using the speech synthesis filter.

3   Description of Speech Synthesis Systems for Evaluation

3.1   *Average Voice Speech Databases Description and Training*

The speech databases required for an average voice HMMs training arose partly from the acoustic data which are used for a speech recognition system in Slovak language and partly from the especially prepared phonetically balanced data used for a speaker dependent HMM-based speech synthesis training. The speech recognition database was algorithmically divided into isolated sentences and individual speakers and it includes together the utterances of the biggest seven male and seven female speakers where each of them contain more than the 450 utterances. The three (two female and one male) phonetically balanced speech databases especially prepared for the purposes of speech synthesis were added to this data. Together, seventeen sub-databases were used for average voice model training with the help of AVSS system [3]. A detailed specification of the obtained average voice speech databases is shown in Table 1.

3.2   *Adaptation Speech Databases Description*

Eight various sized speech databases were used as adaptation input speech data where each of them arose from one phonetically balanced male database. This large single speaker speech database has been carefully recorded under the studio conditions and it consists of the 3667 phonetically balanced Slovak sentences [10]. It was divided into eight sub-databases, where the smallest one consisted of the 29 utterances and as

Table 1 Average voice speech databases specification

Female speech databases

Speaker	1	2	3	4	5	6	7	8	9
No. of utterances	810	545	535	520	570	525	469	4526	330
Duration (min)	138	97	90	105	112	95	79	411	49

Male speech databases

Speaker	1	2	3	4	5	6	7	8	
No. of utterances	517	631	543	767	657	531	571	330	
Duration (min)	89	87	102	140	89	107	86	38	
Total number of sentences: 13377					Total duration: 31 hr 54 min				

Table 2 Adaptation speech sub-databases specification

Name	AD1	AD2	AD3	AD4	AD5	AD6	AD7	AD8	SD	NS
No. of utterances	29	57	115	229	458	917	1834	3667	3667	3667
Duration (min)	3	6	11	21	37	70	141	284	284	284
Note	Adaptation								a	b

[a]speaker dependent database; [b]natural speaker database

the biggest one the entire database was used. Each sub-database between the first and the last one contained twice as many utterances than the previous database. A detailed specification of obtained adaptation sub-databases is shown in Table 2. The table also includes a description of speaker dependent voice, based on the same database, and the natural speaker database which were used for comparison of acquired results.

3.3 Description of All Systems for Evaluation

On the basis of the aforementioned databases, eight new HMM-based speech synthesis systems were created using the speaker adaptation techniques. These systems use previously developed modules for Slovak text analysis together with the proposed language dependent context clustering. A more detailed description of these parts can be found in [11]. For these experiments, a Constrained maximum likelihood linear regression (CMLLR) combined with additional Maximum a posteriori (MAP) adaptation were used. Following table shows a description of the newly created Slovak AVSS systems together with the speaker dependent one, which is used when comparing the results of evaluation (Table 3).

Table 3 Description of all systems for evaluation

	AD1	AD2	AD3	AD4	AD5	AD6	AD7	AD8	SD
Vocod.	AHOcoder	AHOcoder	AHOcoder	AHOcoder	AHOcoder	AHOcoder	AHOcoder	AHOcoder	AHOcoder
Datab. (mins)	3	6	11	21	37	70	141	284	284
Param.	MGC: 39 + lf0: 1, Multiband mixed excitation

4 Evaluation and Results

An evaluation of newly created voices was performed by the objective tests. The evaluation part of Slovak speech male database, which was created by separating of one fifth of the entire corpus, was used for the objective evaluation. Taken together, a spectrum of generated utterances was evaluated by measuring of their mean Mel-cepstral distortion (MCD) and interpolated f0 (fundamental frequency) contours were compared with f0 contour of reference database and speaker dependent voice trained with the help of the same database. The detailed explanation of all experiments and their results will be given in the following section.

A MCD evaluation method represent a distance measure calculated between mel-cepstral coefficients of the reference (or original) and the evaluated speech samples [12]. In these experimentation, 859 recordings of the reference male speaker and the reference speaker dependent voice are considered (the generated speech files can be found here: http://kemt-old.fei.tuke.sk:1025/synteza_web/adapt_eval/). We applied this metric to the mel-cepstral coefficients generated by eight test systems, which are described in Sect. 3.3. In the first case, these coefficients were compared with the reference speaker's mel-cepstral coefficients and subsequently also with the speaker dependent voice's coefficients. The content of generated speech samples of all system and all utterances had the same content as the reference. Finally, the acquired results from all 859 comparisons were averaged to obtain the mean Mel-cepstral distortion of each evaluated system. The results of objective evaluation of the speech synthesis systems with the mean MCD method are shown in Fig. 2.

As we can see, the obtained results show a continuous increase of the quality with the increasing number of adaptation data. The largest qualitative difference is between the AD1 and the AD2 system which may demonstrates the inadequacy of data in the first system. The most interesting result is the minimum decrease of quality between the systems from four to eight. This phenomenon demonstrates the ineffectiveness

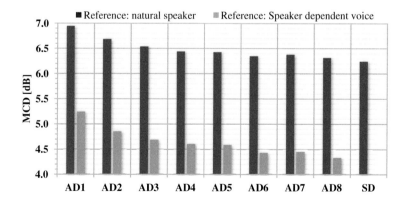

Fig. 2 Mean MCD objective evaluation results

of increasing the number of data at the certain point. The comparison of the adapted systems with the speaker dependent voice as the reference only confirmed the above mentioned conclusions.

The objective evaluation was also made for the fundamental frequency of the generated utterances. In this case, the evaluation part of created corpora was used again for the assessment whether the generated f0 values are correct and appropriate. An evaluation procedure consisted of the fundamental frequency extraction from the male reference database together with the same extraction from the generated utterances of all eight newly created TTS systems. This extraction was performed on each of the 859 recordings of each system and subsequently the alignment was performed with the help of an interpolation to achieve a vector alignment to the same length for the needs of their averaging and comparison. The alignment was followed by the averaging of the values in each sample of acquired vectors what led to obtaining of an average fundamental frequency contour of the generated utterances and the reference samples. The comparison of the aligned and averaged f0 values of all systems together with the reference and speaker dependent voice are shown in Fig. 3. It is apparent that almost all tested systems generate the fundamental frequency of an artificial speech with almost the same values as it is in the reference. The only system with a significantly different f0 contour was the AD1 which again highlights the inadequacy of the adaptation data.

Figure 4 shows the dependency between the mean MCD and the adaptation database size expressed in minutes. As can be seen, it is possible to determine a threshold (between the values from 60 to 70 min), where the adaptation database enlargement already had no significant effect on the output voice quality. The values above this threshold than cause only the quality oscillation.

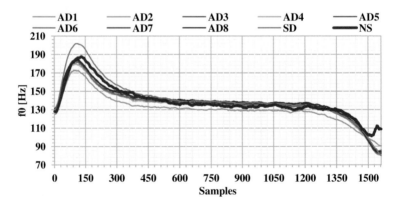

Fig. 3 Comparison of aligned and averaged f0 values

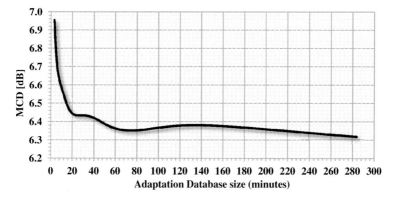

Fig. 4 Dependency between the mean MCD and adaptation database size

5 Conclusions

This paper presents the influence of adaptation database size on the quality of HMM-based synthetic voice based on the large average voice model. The performance of the newly created systems have been evaluated through the objective tests. The above-mentioned tests showed that when the large diverse average model is used it is possible to get quite good synthetic voice with the help of adaptation techniques with the small amount of data. The experiments also showed the existence of the threshold beyond which the increasing size of the database cause only the quality oscillation. Acquired results underline the effectiveness of the model adaptation techniques as such especially when using the AHO-coder which represents one of the most advanced vocoder.

Acknowledgments The research presented in this paper was supported by the Ministry of Education, Science, Research and Sport of the Slovak Republic under the project VEGA 1/0075/15 (50 %) and the Research and Development Operational Program funded by the ERDF under the projects ITMS-26220220182 (50 %).

References

1. Taylor, P.: Text-to-Speech Synthesis. Cambridge University Press, Cambridge (2009)
2. Tokuda, K., Nankaku, Y., Toda, T., Zen, H., Yamagishi, J., Oura, K.: Speech synthesis based on hidden markov models. In: Proceedings of the IEEE, vol. 101, pp. 1234–1252 (2013)
3. Yamagishi, J., Kobayashi, T.: Average-voice-based speech synthesis using HSMM-based speaker adaptation and adaptive training. Trans. Inf. Syst. archive, vol. E90-D, no. 2, pp. 533–543 (2007)
4. Yoshimura, T., Tokuda, K., Masuko, T., Kobayashi, T., Kitamura, T.: Mixed excitation for HMM-based speech synthesis. In: Proceedings of the Eurospeech, pp. 2259–2262 (2001)
5. Erro, D., Moreno, A., Bonafonte, A.: Flexible harmonic/stochastic speech synthesis. In: Proceedings of the 6th ISCA Workshop on Speech Synthesis, pp. 194–199 (2007)

6. Maia, R.S., Toda, T., Zen, H., Nankaku, Y., Tokuda, K.: An excitation model for HMM-based speech synthesis based on residual modeling. In: Proceedings of the 6th ISCA Workshop on Speech Synthesis, pp. 131–136 (2007)
7. Kawahara, H.: Straight, exploitation of the other aspect of vocoder: perceptually isomorphic decomposition of speech sounds. Acoust. Sci. Technol. **27**, 349–353 (2006)
8. Erro, D., Sainz, I., Navas, E., Hernaez, I.: Harmonics plus noise model based vocoder for statistical parametric speech synthesis. J. Sel. Top. Signal Process. **8**, 184–194 (2014)
9. Yamagishi, J., Kobayashi, T., Nakano, Y., Ogata, K., Isogai, J.: Analysis of speaker adaptation algorithms for HMM-based speech synthesis and a constrained SMAPLR adaptation algorithm. Trans. Audio Speech Lang. Process. **17**, 66–83 (2009)
10. Sulír, M., Juhár, J.: Design of an optimal male and female Slovak speech database for HMM-based speech synthesis. In: Proceedings of the Redžúr, pp. 5–9 (2013)
11. Sulír, M., Juhár, J.: Comparison of three vocoding techniques in subjective and objective evaluation tests of Slovak HMM-based TTS systems, in review process (2015)
12. Kubichek, R.: Mel-cepstral distance measure for objective speech quality assessment. In: Proceedings of the Pacific Rim Conference on Communications, pp. 125–128 (1993)

Comparison of Text-Independent Original Speaker Recognition from Emotionally Converted Speech

Jiří Přibil and Anna Přibilová

Abstract The paper describes an application of the classifier based on the Gaussian mixture models (GMM) for reverse identification of the original speaker from the emotionally transformed speech in Czech and Slovak. We investigate whether the identification score given by the GMM classifier depends on the type and the structure of used speech features. Comparison of the results obtained with the sentences in German and English has shown that the structure and the balance of the speech database have influence on the identification accuracy but the used language is not practically important. The evaluation experiments confirmed that the developed text-independent GMM original speaker identifier is functional for the closed-set classification tasks.

Keywords Emotional voice conversion · Speech spectral features · Speech prosodic features · Gaussian mixture model · Original speaker identification

1 Introduction

Expression of positive or negative emotional states in speech, although formerly investigated merely by psychologists, has become a part of a multidisciplinary field of research focused on human-computer interaction [1]. Greater effectiveness of this interaction and dialogue management is achieved by inclusion of emotions in the speech recognition as well as in the speech synthesis. For that reason, the expressive speech synthesis has moved to the centre of attention of speech processing researchers [2] in particular because the improvement of speech naturalness can be achieved by

J. Přibil (✉)
Institute of Measurement Science, SAS, Dúbravská cesta 9,
SK-841 04 Bratislava, Slovakia
e-mail: Jiri.Pribil@savba.sk

A. Přibilová
Faculty of Electrical Engineering & Information Technology, Institute of Electronics and Photonics, Slovak University of Technology, Ilkovičova 3, SK-812 19 Bratislava, Slovakia
e-mail: Anna.Pribilova@stuba.sk

© Springer International Publishing Switzerland 2016   137
A. Esposito et al. (eds.), *Recent Advances in Nonlinear Speech Processing*,
Smart Innovation, Systems and Technologies 48,
DOI 10.1007/978-3-319-28109-4_14

the way of emotional speech style conversion which holds for many languages [3]. In addition, there are increasing demands for realization of more familiar human-computer interfaces [4] using different approaches: development of the text-to-speech (TTS) systems with the expressive speech style production [5], personification of the TTS systems [6], application of the storytelling speaking style for narration of stories for children [7, 8] or in special book reading software for blind users [9], etc. On the other hand, this conversion could not change the original speaker identity in contrast to voice transformation [10]. Subjective listening tests are often used for evaluation of quality, naturalness, and intelligibility of the synthetic speech [11], however, in our case the successfulness of this type of evaluation is very problematic–practically impossible. For this reason, we sought another statistically-based evaluation method that can be used for the original speaker identification from the resynthesized speech with applied emotional speech style conversion.

In the previous work [12] it was verified that the synthetic speech quality can be evaluated by the original speaker identification using text-independent classification in a closed set with the unknown speaker in the set of known speakers [13]. The Gaussian mixture model (GMM) of a speaker [14], providing a probabilistic model of the underlying sounds of a person's voice, is useful for text-independent speaker identification with short duration of speech utterances [15, 16]. The main advantage of this statistical evaluation method is that it needs no human interaction and the obtained results can be compared numerically. The accuracy of identification of the original speaker from the signal generated by the speech synthesizer depends on the used method of synthetic speech production—it means that the changes (errors, deviations from the original signal) are caused only by the chosen method of speech signal parameterization. In the case of emotional speech transformation, much more factors have influence on the original speaker identification accuracy due to the modifications in the spectrum (changed formant positions) and the prosody (changed F0 contour, energy contour, time duration, linear trend, etc.). Identification of the original speaker may be affected by these changes with possible decrease of the final GMM identification accuracy.

Motivation of the work described in this paper was to verify applicability of a GMM-based classifier for identification of the original speaker after applied conversion from neutral to four emotional states (joy, surprise, sadness, and anger) spoken in Czech and Slovak languages [17]. In the comparison experiment, the neutral speech was first emotionally converted and subsequently used for identification of the original speaker. In this way, the impact of emotional conversion upon speaker identification is investigated in our experiment. In addition, the correctness of the GMM identification is analyzed with respect to the setting of the parameters and the choice of the speech features during GMM training. The emotional speech conversion method used in this experiment was originally developed within the framework of previous research [18]. This method is based on non-linear spectral envelope transformation that shifts the first formant to the left and the higher ones to the right for pleasant emotions, and the first formant to the right and higher ones to the left for unpleasant emotions according to the knowledge of psychological and phonetic research [19]. For the speech analysis and resynthesis of the male voice the source-filter model with cepstral parameterization of the vocal tract transfer function [20] was used, the speech signal of the female voice was resynthesized using the harmonic speech model [18, 21]. The databases containing original neutral speech used in our research were: the database of Czech and Slovak stories (CZ&SK) performed by professional actors [17], the Berlin Database of Emotional Speech (Emo-DB) in German language [22] and the Texas Instruments and Massachusetts Institute of Technology (TIMIT) database in English language [23].

2 Description of Used GMM-based Re-identification Method

The GMMs represent a linear combination of multiple Gaussian probability distribution functions of an input data vector [14]. For GMM creation, it is necessary to determine the covariance matrix, the vector of mean values, and the weighting parameters from the input training data. Using the expectation-maximization (EM) iteration algorithm, the maximum likelihood function of GMM is found [14]. The EM algorithm is controlled by the number of used mixtures (N_{GMIX}) and the number of iterations (N_{ITER}); the iteration stops when the difference between the previous and current probabilities fulfills the internal condition or the predetermined maximum number of iterations is reached. In general, the elements of the feature vectors could be correlated [17], so rather a high number of mixtures and a full covariance matrix would be necessary to provide sufficient approximation [24]. On the other hand, the GMM with a diagonal covariance matrix is usually used for speaker identification [14] due to lower computational complexity. The GMM classifier returns the probability (so called *score*) that the tested utterance belongs to the GMM model. In the standard realization of the GMM classifier, the resulting class is given by the maximum overall probability of all obtained scores corresponding to M output classes using the feature vector T from the tested sentences

$$m^* = \arg \max_{1 \leq m \leq M} score\, (T, m) . \tag{1}$$

This relatively simple and robust approach cannot achieve the best recognition accuracy in all cases. In our experiment, a more complex method based on the accumulated score calculation was used for final decision about the classified original speaker. The accumulated score can be expressed by the relation

$$m_{ACC} = \arg \max_{1 \leq i \leq M} \bigcup_{p=1}^{P} \left(m^* (i, p) \equiv i \right), \tag{2}$$

where $m^*(i, p)$ represents the value calculated by (1) for the current pth frame, P is the number of the frames in the sentence, and the union operator represents the occurrence rate of the ith class. Practical realization consists of an experimental one-

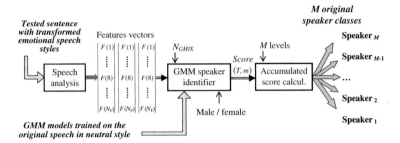

Fig. 1 Block diagram of the developed GMM-based classifier for identification of the original speaker from the converted emotional speech

level structure of the GMM classifier as shown in Fig. 1. Classification is performed for both genders of the voice (male/female) that had been correctly determined in the previous process [17] or set manually. The input feature vectors are processed from the tested sentences with the transformed emotional speech. The speaker identification block uses the GMM models that were created and trained on the data of the feature vectors obtained from the sentences of original speakers in a neutral style. The obtained individual values of *score* (T, m) are further used for calculation of the accumulated score m_{ACC} and depending on the used discrimination level the M output classes of original speakers are finally determined.

The speech signal analysis is performed in the following way: the fundamental frequency F0 is determined from the input sentence after segmentation and weighting. For all speech frames the smoothed spectral envelope and the power spectral density are computed for further processing: determination of spectral and prosodic features. The speech parameter vector of N values can be next processed in two ways:

- the first one uses only one representative statistical value (mean value, standard deviation, median value, etc.) from all N values,
- the second one uses $N - 2K$ representative statistical values computed from a window with the length of $W_{AVER} = 2K + 1$ values around the ith value where $K + 1 \leq i \leq N - K$.

These output feature vectors with the length of N_{FEAT} are stored in the database for direct use in the GMM classification process—see the block diagram in Fig. 2. Spectral features like mel frequency cepstral coefficients (MFCC) together with energy and prosodic parameters are most commonly used in GMM-based speaker identification [25] and emotional voice classification [26]. However, the relative position of formants and formant trajectories [27] can be used as the main indicator for speech classification in voiced parts. The basic spectral features comprising the first two formant positions $F_{1,2}$ [28] with theirs ratios F_1/F_2 and the first four cepstral coefficients c_{1-4} are used here together with the features determined from the spectral envelope: decrease, spread, and centroid. The used supplementary spectral features are: harmonic-to-noise ratio (HNR), spectral flatness (SFM), and spectral entropy (SE). The conversion method includes also the changes in the time

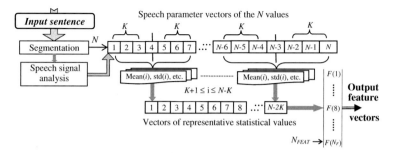

Fig. 2 Block diagram of the determination of the feature vectors from the speech spectral properties and the supra-segmental parameters

duration—lengthening or shortening in dependence on the chosen type of emotional style conversion [29]. As regards the supra-segmental speech properties, the prosodic parameters like differential contour $F0_{DIFF}$ (F0 after subtraction of its mean and the LT removal), zero-crossing rate ($F0_{ZCR}$), jitter, shimmer, etc. were determined.

3 Speech Material and Performed Experiments

The main speech corpus used for GMM creation, training, and testing consists of 89 sentences uttered by 5 male speakers and 79 sentences uttered by 5 female speakers with duration from 0.5 to 8.5 s, resampled at 16 kHz. The sentences from the CZ&SK database in a neutral style were subsequently converted to four emotional styles: "joy", "joyous surprise", "sadness", and "anger". The structure of the speech corpus, the composition of the speakers, and the record time durations in the Czech and Slovak speech database are relatively similar to those in the well-known Berlin Database of Emotional Speech [22] in German language that is the only public and free database among several emotional speech databases which are either private or commercially available or public with licence fee [30]. The principal reason for the use of the Emo-DB was that it can be taken as a reference for the final comparison experiment. The Emo-DB speech database consists of the set of sentences with the same contents expressed in seven emotional styles: "neutral", "joy", "sadness", "boredom", "fear", "resistance", and "anger". For processing in this experiment, only neutral sentences uttered by 5 male and 5 female speakers with duration from 1.5 to 8.5 s were extracted from the whole speech corpus (100 sentences altogether). The TIMIT speech database in English (having been collected since 1990 and consisting of 630 speakers in total, 438 males and 192 females) is often used for GMM recognition/identification [23] as well as in more complex comparisons [31]. In spite of the fact that the TIMIT database consists of the sentences in a neutral style only, we take sentences from this database as the second reference for comparison with the results using the Czech and Slovak speech database. In a similar way as in the case of the Emo-DB, we chose $100 + 100$ sentences by $5 + 5$ male/female speakers with duration from 1.3 to 6.4 s, sampled at 16 kHz.

The frame length for the spectral analysis depends on the mean pitch period of the processed signal. The processed speech material originates from speakers with different mean F0 values, so different parameter settings for analysis—frame (window) length and window overlapping—were applied. The F0 values (for pitch contour calculation) were determined by the autocorrelation analysis method with the pitch-period correction from the cepstrum using the experimentally chosen pitch ranges of $55 \div 250$ Hz for the male voices and $105 \div 350$ Hz for the female ones. The input feature vector length was set to $N_{FEAT} = 16$ and five feature sets were created as shown in Table 1.

For the experiment with GMM recognition of the original speaker, the analysis and comparison was aimed at investigation of:

- influence of different types of speech parameters used in the five sets of the input feature vectors—see the numerical results for all four types of transformed emotions in Table 2 and the 3D confusion matrices for the male and the female voices in Fig. 3,
- influence of the initial parameter during the GMM creation on the resulting identification score: the number of applied mixtures of the Gaussian probability density functions $N_{GMIX} = \{8, 16, 32, 64, 128, 256,$ and $512\}$—see the summarized mean values in Table 3,
- influence of the chosen window length W_{AVER} on computation of representative statistical values used in the input feature vectors for GMM training and testing; the evaluation done for $K = \{1, 3, 5, 10, 15, 20\}$—see detailed results in Fig. 4,

Table 1 Structure of the used feature sets for the GMM re-identification

Set	Feature type	Statistical value
P0	{HNR, spectral decrease, centroid, SFM, SE, F0$_{DIFF}$, jitter, and shimmer}	{min, rel. max, min, mean, std, median}
P1	{spectral spread, decrease, centroid, SFM, HNR, F0$_{DIFF}$, F0$_{ZCR}$, jitter, and shimmer}	{mean, median, std, rel. max, min, max}
P2	{F_1/F_2, spectral decrease, centroid, HNR, SFM, SE, F0$_{DIFF}$, jitter, and shimmer}	{mean, std, median}
P3	{$F_1, F_2, F_1/F_2$, spectral decrease, HNR, SFM, SE, F0$_{DIFF}$, jitter, and shimmer}	{skewness, kurtosis, std, mean, median, rel. max, max}
P4	{$c_1 - c_4$, spectral decrease, centroid, SFM, SE, F0$_{DIFF}$, jitter, and shimmer}	{skewness, mean, std, median}

Table 2 Comparison of the mean original speaker GMM recognition accuracy in [%] including values of the standard deviation (in parentheses), documenting influence of the used type of the feature vector; results for sentences with all four types of transformed emotions altogether

Voice/feature set	P0	P1	P2	P3	P4
Male	93.3 (4.2)	91.7 (7.1)	91.7 (6.5)	76.6 (14)	90.2 (6.5)
Female	93.9 (3.8)	94.6 (5.7)	92.8 (5.7)	68.1 (20)	93.9 (3.8)
Total	**93.65**	**93.15**	**92.25**	**72.35**	**92.05**

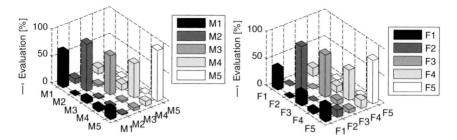

Fig. 3 3-D representation of confusion matrices of the GMM original speaker recognition for the speech with all four transformed emotions altogether: male (*left*) and female (*right*) voices, feature set P3 (the worst one)

Table 3 Mean values of the original speaker recognition accuracy in [%] including the standard deviation (in parentheses) for sentences with transformed emotions altogether depending on the used number of mixtures N_{GMIX}

Voice/N_{GMIX}	5[a]	8	16	32	64	128	256	512
Male	77.8 (6.5)	88.7 (9.4)	90.8 (8.8)	93.6 (4.2)	93.3 (4.2)	91.0 (7.1)	84.9 (16.2)	78.2 (29.8)
Female	79.5 (18.2)	69.5 (31)	83.4 (16.3)	92.6 (7.3)	94.0 (3.8)	94.8 (2.6)	94.3 (5.5)	84 (11.5)
Total	**78.7**	**77.3**	**87.2**	**93.1**	**93.7**	**92.9**	**89.6**	**81.1**

[a]Used the feature set P0et

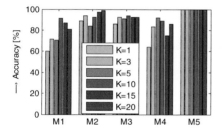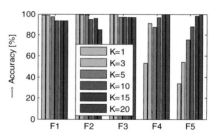

Fig. 4 Influence of the used window length ($2K + 1$) for the speech feature determination on the GMM original speaker recognition accuracy; results for all four types of the transformed emotions: male (*left*) and female (*right*)

- analysis of the GMM recognition accuracy for each of the four transformed emotional styles—see the summary results in Table 4,
- comparison of the computational complexity: CPU times for the GMM creation and training phase as well as mean values of the original speaker classification accuracy for different number of used mixtures; summarized for all four types of transformed emotions and both voices presented by Table 5,
- final analysis of the original speaker identification accuracy for utterances from CZ&SK database in comparison with the results obtained for utterances from Emo DB and TIMIT—see summary values of the achieved accuracy in Table 6.

Table 4 Summarized mean values of the GMM recognition accuracy in [%] including values of the standard deviation (in parentheses) sorted by the type of the transformed emotional style

Emotion/speaker	Joy	Surprise	Sadness	Anger
Male	97.8 (4.9)	100 (0)	89.6 (9.6)	89.4 (11)
Female	97.6 (3.0)	98.7 (2.7)	89.7 (12)	89.3 (7.6)
Total	**97.70**	**99.35**	**89.65**	**88.35**

Table 5 Comparison of the computational complexity (CPU time in [s]) for different number of used mixtures; summarized for all transformed emotions and both genders

Phase/N_{GMIX}	5[a]	8	16	32	64	128	256	512
Creation and training[b]	0.58	11.8	20.5	43.3	87.5	179	363	861
Identification[c]	0.25 (15.4)	0.64 (37.6)	0.65 (38.4)	0.71 (43.9)	0.74 (43.1)	0.93 (53.8)	1.17 (67.7)	1.69 (95.4)
Total time	**0.825**	**12.44**	**21.15**	**44.01**	**88.24**	**179.9**	**364.2**	**862.7**

[a]The used feature set P0et
[b]Summary values for all transformed emotional speech styles in both genders (5 + 5 models)
[c]Mean values per sentence including the standard deviation values in [ms] (in parentheses)

Table 6 Final comparison of the GMM recognition accuracy in [%] including the values of the standard deviation (in parentheses) depending on the used speech database

Voice/database	CZ&SK[a]	CZ&SK	Emo-DB	TIMIT
Male	77.8 (6.5)	93.3 (4.2)	85.5 (8.9)	95.5 (3.7)
Female	79.5 (18.2)	94.0 (3.8)	84.0 (11.7)	97.5 (2.5)
Total	**78.7**	**93.6**	**84.8**	**96.5**

[a]The feature set P0et and the simple classification method was used

If not defined otherwise, the setting used for the first three comparisons was: feature set P0, $N_{GMIX} = 64$, $N_{ITER} = 1000$, $K = 10$. In the final analysis, the results were compared with a simple method of calculation of representative statistical values in the input vectors (one value for every sentence) and using a simple classification method from the score determined by (1). In this approach, the feature set P0et having the same structure as the set P0, and setting of $N_{GMIX} = 5$, $N_{ITER} = 500$ was applied. The obtained results are presented for visual comparison using the graphical form (the confusion matrices and/or the bar graphs of the identification accuracy in [%]) as well as numerical matching of the mean values stored in the tables. The original speaker identification accuracy was calculated from X_A sentences with correctly identified original speaker and the total number N_U of the tested sentences as $(X_A/N_U)*100$ [%]. The values in the confusion matrices were calculated in a similar way.

The computational complexity was tested on the PC with the following configuration: processor Intel(R) i3-2120 at 3.30 GHz, 8 GB RAM, and Windows 7 professional OS. To determine the spectral features and the prosodic parameters, the elementary functions from Matlab ver. 2010b environment with the help of "Signal Processing Toolbox" and "Statistics Toolbox" were applied. The basic functions from the Ian T. Nabney "Netlab" pattern analysis toolbox [32] were used for implementation of the GMM functions.

4 Discussion of Obtained Results

Results of the first analysis have shown that proper selection of the input features for GMM evaluation is very important. Contrary to our expectations, for the extended sets P3–P4 comprising the formants and the cepstral coefficients, there was no increase in the original speaker identification accuracy. In the case of P3 containing the first two formant positions and their ratios, the results were the worst—the achieved recognition accuracy of the female speakers was even lower than 50 % (see Fig. 3). For GMM original speaker identification from the emotionally converted sentences, the best results were obtained with P0 containing a mix of spectral and basic prosodic speech features; therefore, it was used in the further experiments.

In general, it holds that higher number of Gaussian mixtures can increase the recognition accuracy when the short speech signal is evaluated (with duration up to one second [24]). For this reason, at first, we carried out the analysis of influence of the used number of GMM mixtures in a larger interval spanning from 8 to 512 mixtures. Numerical matching of obtained results in Table 3 shows that a relatively higher improvement was observed in the summary mean recognition accuracy of 93/94 % (for male/female voices—in the best case of $N_{GMIX} = 64$) compared with 89/70 % accuracy for the minimum number of 8 mixtures. The applied number of mixtures has a great influence on the computational complexity (the measured CPU time) for creation and training of the GMM models but it has only a little impact on duration of the identification phase—see values in Table 5. The use of the maximum number of 512 mixtures causes increase of the CPU time more than 10 times when compared with 64 mixtures (and approximately 70 times higher CPU time than for 8 mixtures). Moreover, as the maximum value of $N_{GMIX} = 512$ does not bring the best results of the recognition accuracy, for next processing the setting $N_{GMIX} = 64$ was applied. On the other hand, choice of the number of iterations N_{ITER} has not great weight when its order is about hundreds; the optimum value is about one thousand.

The subsequently performed analysis of the influence of different window lengths W_{AVER} for the speech feature determination has shown a local maximum of the achieved identification accuracy lying in the interval of $K = 10 \div 15$ with slight differences for the male and the female voices (see bar-graphs in Fig. 4). Therefore, the value $K = 10 (W_{AVER} = 21)$ was chosen for next experiments. The detailed results per emotions (see Table 4) are in accordance with the degree of the applied changes during the emotional style transformation: the best score for the style of the "joyous surprise" (99 %), the worst score for the style of the "anger" (88 %). When the feature set P0ct and the simple classification method are used, the obtained identification accuracy is much lower—about 78 % as shown in the first column in Table 6.

The final comparison with Emo-DB and TIMIT shows heavy dependence of the achieved identification accuracy on the used speech database for the GMM training and testing. In the case of analysis using the sentences from the CZ&SK we obtained slightly better results than for the utterances originated from the Emo-DB. It might be caused by the selection of only sentences in a neutral speech style from the Emo-DB for next processing, so from the statistical point of view the used speech corpus was not finally balanced. It seems to be a cause of higher dispersion of the results for each of the female speakers in Emo-DB as it can be seen in Table 4. The TIMIT database gives higher re-identification accuracy for male voices (up to 96 % in comparison with 93 % for CZ&SK) as well as for female voices (97 % vs. 94 % for CZ&SK). These results could depend on total neutrality of the TIMIT database while CZ&SK and Emo-DB were primarily constructed for the experiments with emotional speech.

5   Conclusion

The obtained results of the basic original speaker identification experiment corresponded with the proposed working hypothesis about lowering of the achieved GMM identification score for greater range of the performed spectral and prosodic modifications of the emotionally transformed speech. Our approach of applying the GMM for the original speaker identification from the speech after conversion from neutral to four emotional states uses atypically relatively a low number of the specific speech features (unlike the mostly used MFCC coefficients)—energy parameters, low-pass intensity, or high-pass intensity—as well as the number of mixtures for the GMM model creation, training, and classification. In difference of our previous works [12, 17, 33], for the final speaker re-identification the approach based on cumulative score was applied. The last specificity of this research and comparison lies in the fact that the basic speech database used for experiments was composed of the sentences uttered by the original Czech and Slovak speakers. The parameters used for emotional speech style conversion were also determined using the research results aimed at statistical analysis of Czech and Slovak emotional speech. Therefore, it is hard to compare the obtained results directly with the current state of the art published by other authors for different languages. Yet, the basic comparison of the original speaker identification using the German (Emo-DB) and English (TIMIT) speech has shown that the identification accuracy depends also on the structure and the balance of the used speech database but the influence of the language of the utterance is not very important.

The next aim of finding the best (optimal) structure of the input feature set for GMM original speaker recognition was fulfilled, too. Some types of speech features are not sufficient for this identification task, especially those based on formant frequencies—see the worst results for the set P3. On the other hand, the choice of the type of the used statistical representative value is not substantial. The discovered local maximum of the used number of the GMM mixtures corresponds with reasonable requirement on computational complexity as documented by the results in Tables 3 and 5. In the speech features determination phase, the influence of the window length on the computational complexity was only minimal, but the correct setting had significant influence on the original speaker identification accuracy.

The currently developed method uses the diagonal covariance matrix of GMM models [12]. Therefore, in near future we will compare this approach with the other ones which use the full covariance matrix or the probabilistic PCA (Principal Component Analysis) [34] although at the expense of higher computational complexity. Further, we plan also to compare identification using the GMM approach and the other favoured methods such as SVM [35].

Acknowledgments This work has been supported by the and VEGA 1/0090/16 Grant Agency of the Slovak Academy of Sciences (VEGA 2/0013/14) and the Ministry of Education of the Slovak Republic (KEGA 022STU-4/2014).

References

1. Skowron, M., Rank, S., Swiderska, A., Küster, D., Kappas, A.: Applying a text-based affective dialogue system in psychological research: case studies on the effects of system behaviour, interaction context and social exclusion. Cogn. Comput. p. 20 (2014), doi:10.1007/s12559-014-9271-2
2. Maia, R., Akamine, M.: On the impact of excitation and spectral parameters for expressive statistical parametric speech synthesis. Comput. Speech Lang. **28**(5), 1209–1232 (2014)
3. Riviello, M.T., Chetouani, M., Cohen, D., Esposito, A.: On the perception of emotional "voices": a cross-cultural comparison among American, French and Italian subjects. In: Esposito, A., Vinciarelli, A., Vicsi, K., Pelachaud, C., Nijholt, A. (eds.) Analysis of Verbal and Nonverbal Communication and Enactment: The Processing Issues. LNCS, vol. 6800, pp. 368–377. Springer, Berlin (2011)
4. Yun, S., Lee, Y.J., Kim, S.H.: Multilingual speech-to-speech translation system for mobile consumer devices. IEEE Trans. Consum. Electron. **60**(3), 508–516 (2014)
5. Přibil, J., Přibilová, A.: Application of expressive speech in TTS System with cepstral description. In: Esposito, A., Bourbakis, N., Avouris, N., Hatrzilygeroudis, I. (eds.) Verbal and Nonverbal Features of Human-Human and Human-Machine Interaction. LNAI, vol. 5042, pp. 201–213. Springer, Berlin (2008)
6. Hanzlíček, Z., Matoušek, J., Tihelka, D.: First experiments on text-to-speech system personification. In: Matoušek, V., Mautner, P. (eds.) Text, Speech, and Dialogue 2009. LNCS, vol. 5729, pp. 186–193. Springer, Berlin (2009)
7. Lee, H.J.: Fairy tale storytelling system: using both prosody and text for emotional speech synthesis. In: Lee, G., Howard, D., Ślogonezak, D., Hong, Y.S. (eds.) Convergence and Hybrid Information Technology. Communications in Computer and Information Science, vol. 310, pp. 317–324. Springer, Berlin (2012)
8. Alcantara, J.A., Lu, L.P., Magno, J.K., Soriano, Z., Ong, E., Resurreccion, R.: Emotional narration of children's stories. In: Nishizaki, S.Y., Numao, M., Caro, J., Suarez, M.T. (eds.) Theory and Practice of Computation. Proceedings in Information and Communication Technology, vol. 5, pp. 1–14. Springer, Japan (2012)
9. Přibil, J., Přibilová, A.: Czech TTS engine for Braille pen device based on pocket PC platform. In: Vích, R. (ed.) Proceedings of the 16th Conference Electronic Speech Signal Processing ESSP'05 joined with the 15th Czech-German Workshop Speech Processing, pp. 402–408 (2005)

10. Erro, D., Alonso, A., Serrano, L., Navas, E., Hernaez, I.: Interpretable parametric voice conver-
    sion functions based on Gaussian mixture models and constrained transformations. Comput.
    Speech Lang. **30**(1), 3–15 (2015)
11. Tihelka, D., Matoušek, J., Kala, J.: Quality deterioration factors in unit selection speech syn-
    thesis. In: Matoušek, V., Mautner, P. (eds.) Text, Speech, and Dialogue 2007. LNAI, vol. 4629,
    pp. 508–515. Springer, Berlin (2007)
12. Přibil, J., Přibilová, A., Matoušek, J.: GMM classification of TTS synthesis: Identification of
    original speaker's voice. In: Sojka, P., Horák, A., Kopeček, I., Pala, K. (eds.) Text, Speech, and
    Dialogue. LNAI, vol. 8655, pp. 365–373. Springer, Cham (2014)
13. Shahin, I.: Speaker identification in emotional talking environments based on CSPHMM2s.
    Eng. Appl. Artif. Intell. **26**(7), 1652–1659 (2013)
14. Reynolds, D.A., Rose, R.C.: Robust text-independent speaker identification using Gaussian
    mixture speaker models. IEEE Trans. Speech Audio Process. **3**(1), 72–83 (1995)
15. Ajmera, P.K., Jadhav, D.V., Holambe, R.S.: Text-independent speaker identification using
    Radon and discrete cosine transforms based features from speech spectrogram. Pattern Recog-
    nit. **44**(10–11), 2749–2759 (2011)
16. Jawarkar, N.P., Holambe, R.S., Basu, T.K.: Text-independent speaker identification in emo-
    tional environments: a classifier fusion approach. In: Sambath, S., Zhu, E. (eds.) Frontiers in
    Computer Education. AISC, vol. 133, pp. 569–576. Springer, Berlin (2012)
17. Přibil, J., Přibilová, A.: Evaluation of influence of spectral and prosodic features on GMM clas-
    sification of Czech and Slovak emotional speech. EURASIP J. Audio Speech Music Process.
    2013(8), 1–22 (2013)
18. Přibilová, A., Přibil, J.: Harmonic model for female voice emotional synthesis. In: Fierrez, J.,
    et al. (eds.) Biometric ID Management and Multimodal Communication. LNCS, vol. 5707, pp.
    41–48. Springer, Berlin (2009)
19. Scherer, K.R.: Vocal communication of emotion: a review of research paradigms. Speech
    Commun. **40**(1–2), 227–256 (2003)
20. Vích, R.: Cepstral speech model, Padé approximation, excitation, and gain matching in cepstral
    speech synthesis. In: Proceedings of the 15th Biennial EURASIP Conference Biosignal 2000,
    pp. 77–82. Brno, Czech Republic (2000)
21. Madlová, A.: Autoregressive and cepstral parametrization in harmonic speech modelling. J.
    Electr. Eng. **53**(1–2), 46–49 (2002)
22. Burkhardt, F., Paeschke, A., Rolfes, M., Sendlmeier, W., Weiss, B.: A database of German emo-
    tional speech. In: Proceedings of the INTERSPEECH 2005, pp. 1517–1520. Lisbon, Portugal
    (2005)
23. Lopes, C., Perdigão, F.: Phoneme recognition on the TIMIT database. In: I. Ipšić (ed.) Speech
    Technologies, InTech (2011). doi:10.5772/17600
24. Dileep, A.D., Sekhar, CCh.: Class-specific GMM based intermediate matching kernel for clas-
    sification of varying length patterns of long duration speech using support vector machines.
    Speech Commun. **57**, 126–143 (2014)
25. Zhao, X., Wang, D.: Analysing noise robustness of MFCC and GFCC features in speaker
    identification. In: Proceedings of the IEEE International Conference on acoustics, Speech and
    Signal Processing (ICASSP), pp. 7204–7208 (2013)
26. Ooi, C.S., Seng, K.P., Ang, L.M., Chew, L.W.: A new approach of audio emotion recognition.
    Expert Syst. Appl. **41**(13), 5858–5869 (2014)
27. Gharavian, D., Sheikhan, M., Ashoftedel, F.: Emotion recognition improvement using normal-
    ized formant supplementary features by hybrid of DTW-MLP-GMM model. Neural Comput.
    Appl. **22**(6), 1181–1191 (2013)
28. Stanek, M., Sigmund, M.: Comparison of speaker individuality in triangle areas of plane for-
    mant spaces. In: Proceedings of the 24th International Conference Radioelektronika, Bratislava
    2014, p. 4 (2014). doi:10.1109/Radioelek.2014.6828439
29. Wu, C.H., Hsia, C.C., Lee, C.H., Lin, M.C.: Hierarchical prosody conversion using regression-
    based clustering for emotional speech synthesis. IEEE Trans. Audio Speech Lang. Process.
    18(6), 1394–1405 (2010)

30. Sezgin, M.C., Gunsel, B., Kurt, G.K.: Perceptual audio features for emotion detection. EURASIP J. Audio Speech Music Process. **2012**(16) (2012). http://asmp.eurasipjournals.com/2012/1/16
31. Tóth, L., Grósz, T.: A Comparison of deep neural network training methods for large vocabulary speech recognition. In: Habernal, I., Matoušek, V. (eds.) Text, Speech and Dialogue. LNAI, vol. 8082, pp. 36–43. Springer, Berlin (2013)
32. Nabney, I.T.: Netlab Pattern Analysis Toolbox (1996-2001). Retrieved 16 February 2012, from http://www.mathworks.com/matlabcentral/fileexchange/2654-netlab
33. Přibil, J., Přibilová, A.: GMM-Based evaluation of emotional style transformation in Czech and Slovak. Cogn. Comput. p. 11 (2014). doi:10.1007/s12559-014-9283-y
34. Zhao, J., Jiang, Q.: Probabilistic PCA for t-distributions. Neurocomputing **69**(16–18), 2217–2226 (2006)
35. Staroniewicz, P. Majewski, W.: SVM based text-dependent speaker identification for large set of voices. In: Proceedings of the 12th European Signal Processing Conference, EUSIPCO 2004, pp. 333–336. Vienna, Austria (2004)

An Analysis of Shallow and Deep Representations of Speech Based on Unsupervised Classification of Isolated Words

Giampiero Salvi

Abstract We analyse the properties of shallow and deep representations of speech. Mel frequency cepstral coefficients (MFCC) are compared to representations learned by a four layer Deep Belief Network (DBN) in terms of discriminative power and invariance to irrelevant factors such as speaker identity or gender. To avoid the influence of supervised statistical modelling, an unsupervised isolated word classification task is used for the comparison. The deep representations are also obtained with unsupervised training (no back-propagation pass is performed). The results show that DBN features provide a more concise clustering and higher match between clusters and word categories in terms of adjusted Rand score. Some of the confusions present with the MFCC features are, however, retained even with the DBN features.

Keywords Deep learning · Representations · Hierarchical clustering

1 Introduction

Since the introduction of a fast learning procedure for Deep Belief Networks (DBNs) [6], deep learning has provided state-of-the-art performance in many areas of Machine Learning [3, 9], for an extensive review of the applications of these methods see [5]. The paradigm is based on pre-training a DBN, that is, a stack of Restricted Boltzmann Machines (RBMs), on large amounts of unlabelled data and using its parameters to initialise a discriminative Deep Neural Network (DNN) that is trained on small and labelled data sets. This paradigm has become the standard way of building Automatic Speech Recognition (ASR) systems as well [7]. However, the feature representations extracted by these deep models have only partly been studied. In many of those studies, the representations are analysed in relation to the

G. Salvi (✉)
Department for Speech, Music and Hearing, School of Computer Science
and Communication, KTH, Stockholm, Sweden
e-mail: giampi@kth.se
URL: http://www.speech.kth.se/~giampi/

© Springer International Publishing Switzerland 2016
A. Esposito et al. (eds.), *Recent Advances in Nonlinear Speech Processing*,
Smart Innovation, Systems and Technologies 48,
DOI 10.1007/978-3-319-28109-4_15

labels used for the supervised part of the training. In [4], e.g., the authors investigate if adding the output of all the DNN layers can improve speech recognition. Their conclusion is that the second layer from the top still improves results whereas layers that are closer to the input do not carry useful information.

More extensive analyses have been performed with Convolutional Neural Networks, where it is often possible to interpret the meaning of the weights, at least for the first layers that are closer to the signal representation. This was done in the area of Computer Vision (e.g. [11, 14]), but, recently also for Automatic Speech Recognition, [13]. We are not aware of careful analyses of the representations learned by DBNs/DNNs.

In this paper we investigate how deep features relate to shallow features when no supervision is used. For this reason we trained a DBN on MFCC frames from the TIDIGITS database and we compare the word level clustering that can be obtained with these features as opposed to the same clustering based on the MFCC features alone. The data used for clustering is the subset of isolated digits from the database, and the pairwise distance between speech examples is computed with the help of Dynamic Time Warping (DTW).

2   Method

The objective of the method is to establish how discriminative and robust are shallow and deep features when used to compare utterances of isolated words. We used Dynamic Time Warping (DTW) to compare different utterances, with L1 norm of the local distances between feature vectors. The features consisted of simple MFCC vectors in the shallow case and the output of the 4th layer of an DBN in the deep case.

Hierarchical clustering was performed with complete linkage, based on the pairwise distance between utterances. The validity of the clustering was estimated with respect to the true digits. We used for this purpose the adjusted Rand index [8], that measures the similarity between two partitions disregarding the actual labels used (which are arbitrary). If S is a set of n elements (the spoken utterances in our case), and X and Y are two partitions of S of sizes r and s respectively with, possibly $r \neq s$, then the Rand index [12] is defined as:

$$R = \frac{a+b}{\binom{n}{2}},$$

where, a is the number of pairs of elements in S that are in the same set in X and in the same set in Y, and b is the number of pairs of elements in S that are in different sets in X and in different sets in Y. The adjusted Rand index corrects the Rand index for chance.

This measure was used both to determine an optimal level at which the clustering tree should be cut, but also to compare between the partitions obtained with different methods.

3 Data

We used the TIDIGITS database for this study [10]. The database contains sequences of digits spoken by American English speakers. Eleven words are contained in the database because the digit "0" can be pronounced both with the words "oh" or "zero". The database is divided into training and test set. The training set, that was used for this study, contains 112 speakers (57 women and 55 men), each uttering 77 sequences of digits. Of these, 22 are isolated digits (two repetitions of each digit word). In all cases, the clustering is performed on the isolated digits, whereas the sequences are used for training the DBN. In total, 6159 utterances (1,267,952 frames) were used for training, and 2464 utterances (239,440 frames) for clustering.

4 Experiments

The MFCC features were computed over windows of 20 ms with 10 ms steps and 512 point FFT length. Thirteen MFCC coefficients were used including C_0–C_{12}. No delta coefficients were used for this study, neither for the clustering nor as input to the DBN model. The DBN includes four hidden layers of 1024 Bernoulli nodes each. The input layer consists of 13 Gaussian nodes. The DBN was trained using the "pdnn" package [1], based on the Theano toolkit [2], with the following parameters: epochs per layer: 10, batch size: 128, learning rate (Bernoulli-Bernoulli): 0.08, learning rate (Gaussian-Bernoulli): 0.005, initial momentum 0.5 (used for 5 epochs), final momentum: 0.9. The pre-training took about 2.5 h on a GeForce GTX TITAN GPU.

The original MFCC vectors and the activity at the 4th layer at the DBN were then used to compute a 2464 × 2464 distance matrix between each pair of utterances in the test set (isolated digit files). The computation took in each case about 24 h on 24×Intel(R) Xeon(R) CPU X5660 @ 2.80GHz cores.

5 Results

Figure 1 shows the full dendrogram for the clustering obtained with MFCC and DBN features. Given that the tree has 2464 leaves, it is not possible to display the labels associated with them. The figure is included to give a global idea of how the clustering looks like.

Figure 2 displays the Rand score as a function of the number of clusters (and in turn as a function of the distance level at which the dendrogram of Fig. 1 is cut). The graph has been restricted to up to 200 clusters for clarity of exposition. Between

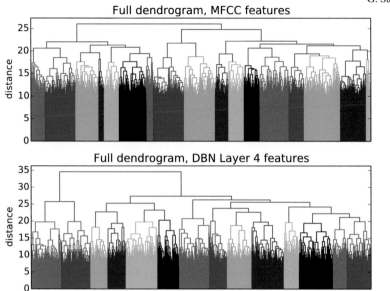

Fig. 1 Dendrogram for the clustering obtained with MFCC features (*top*) and DBN layer 4 features (*bottom*)

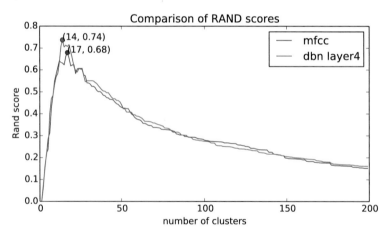

Fig. 2 Adjusted rand score comparing the clustering obtained at different levels of the hierarchy to the actual digits. MFCC features, and DBN layer 4 features

Table 1 Details on the adjusted rand score around the optimum

Adjusted rand index	Number of clusters									
	10	11	12	13	14	15	16	17	18	19
MFCC	0.58	0.61	0.64	0.63	0.63	0.62	0.66	**0.68**	0.65	0.62
DBN4	0.56	0.58	0.63	0.71	**0.74**	0.70	0.71	0.71	0.71	0.68

See also Fig. 2

Table 2 Description of the clusters obtained with the MFCC features (*top*) and DBN features (*bottom*)

MFCC features

Cluster id	Max digit	Digits											Gender	
		o	z	1	2	3	4	5	6	7	8	9	woman	man
2	o	116	0	0	0	0	110	0	0	0	0	0	224	2
10	o	98	0	0	0	0	0	0	0	0	0	0	0	98
1	z	0	104	0	0	0	0	0	0	0	0	0	104	0
12	z	0	110	0	0	0	0	0	0	0	0	0	2	108
3	1	0	0	89	0	0	0	0	0	0	0	77	166	0
13	1	0	0	112	0	0	0	2	0	2	0	100	22	194
15	2	6	4	4	216	6	4	2	0	14	2	4	116	146
5	3	0	0	0	0	108	0	0	0	0	0	0	108	0
14	3	0	0	0	0	110	0	0	0	0	0	0	4	106
11	4	2	0	0	0	0	110	0	0	0	0	0	4	108
6	5	0	0	0	0	0	0	192	0	0	0	2	98	96
7	5	2	4	7	0	0	0	28	0	0	0	10	19	32
8	6	0	0	0	0	0	0	0	224	0	0	0	114	110
16	7	0	0	0	0	0	0	0	0	186	0	0	88	98
17	7	0	2	2	6	0	0	0	0	22	0	2	34	0
9	8	0	0	0	2	0	0	0	0	0	222	0	112	112
4	9	0	0	10	0	0	0	0	0	0	0	29	39	0

DBN layer 4 features

Cluster id	Max digit	Digits											Gender	
		o	z	1	2	3	4	5	6	7	8	9	woman	man
6	o	98	7	0	2	0	0	0	0	44	0	0	149	2
14	o	120	0	45	0	2	2	0	0	0	0	4	65	108
11	z	0	110	0	0	0	0	0	0	0	0	0	106	4
13	z	0	106	0	0	0	0	0	0	0	0	0	0	106
10	2	0	0	0	220	0	0	0	2	2	11	0	116	119
3	3	0	0	32	0	90	0	0	0	0	0	0	120	2
8	3	0	0	0	0	128	0	0	0	0	0	0	20	108
7	4	2	0	0	0	0	222	0	0	0	0	0	112	112
5	5	4	1	0	0	4	0	218	0	0	0	1	122	106
1	6	0	0	0	0	0	0	0	222	0	0	0	112	110
9	7	0	0	0	0	0	0	0	0	178	0	0	68	110
2	8	0	0	0	2	0	0	0	0	0	213	0	114	101
4	9	0	0	31	0	0	0	0	0	0	0	101	132	0
12	9	0	0	116	0	0	0	6	0	0	0	118	18	222

Clusters are ordered by the majority count of digits in each cluster (max digit). The number of clusters is 17 for MFCC features and 14 for DBN features. The number of examples of each digit for each cluster is reported as well as the number of examples spoken by female or male speakers

200 and 2464 clusters the Rand score continues to decrease monotonically both for MFCC and DBN features. From the figure it can be observed that the optimal number of clusters in this case is 17 for MFCC features and 14 for DBN features. The maximum score obtained is higher for DBN features (0.74) than for the MFCC features (0.68). Table 1 shows the Rand score in more details around the optimum. From the table we can see that if we force the number of clusters to be equal to the number of words (11), then the Rand score for the MFCC features is slightly higher than that for DBN features (0.61 vs. 0.58).

We can also see that the Rand score for the DBN4 features is higher than that for the MFCC features even for 17 clusters, where MFCCs perform at best.

Table 2 shows the relation between clusters and actual digits. The first column shows the arbitrary cluster index. The second column shows the digit that corresponds to the majority of examples in that cluster. Two different clustering results are shown: results based on MFCC features (top table) comprise 17 clusters whereas results based on DBN features have only 14 clusters. Both for MFCC and DBN features, it can be observed that in a few cases, more than one cluster corresponds to a specific digit. In some cases, such as digit 'oh', 'zero' and 'three' with MFCC features, it is clear that the two associated clusters correspond to the digit as spoken by female or male speakers respectively. In other cases, such as digit 'five', this distinction cannot be observed. Sometimes, the same cluster includes a nearly equal number of examples from two digits. This is the case, e.g., for digits "oh" and "four" with MFCC features (but not with DBN features), and digits "one" and "nine" for both MFCC and DBN features.

In a few cases the clusters obtained with DBN features are clearly superior. This is the case for the digits 'four', 'five' and 'seven', that obtain a single clusters with DBN features. In other cases the advantage of using DBN features is less evident. For example, there are two clusters for 'nine' that include many examples of 'one'.

6 Conclusions

We studied the properties of MFCC (shallow) and DBN (deep) features for an isolated word clustering task. We observed that DBN features result in fewer and more concise clusters and a higher match with the true identity of the words (14 clusters and 0.74 Rand score for the DBN features; 17 clusters and 0.68 Rand score for MFCC features). However, some of the confusions that we could observe with MFCC features are still present with DBN features, and the latter are still dependent on speaker identity, for example gender. Furthermore, if we force the number of clusters to be equal to the number of words, the difference in clustering performance between MFCC and DBN features is reduced.

Future work will compare the clustering performance using each of the DBN layers. Furthermore, it would be interesting to study the effect of supervised training, for example by adding a softmax layer to the DBN with phonetic classes (or, more precisely HMM states) and retraining the model with back-propagation.

Acknowledgments The GeForce GTX TITAN used for this research was donated by the NVIDIA Corporation.

References

1. https://github.com/yajiemiao/pdnn
2. Bastien, F., Lamblin, P., Pascanu, R., Bergstra, J., Goodfellow, I.J., Bergeron, A., Bouchard, N., Bengio, Y.: Theano: new features and speed improvements. In: Deep Learning and Unsupervised Feature Learning NIPS 2012 Workshop (2012)
3. Ciresan, D., Meier, U., Schmidhuber, J.: Multi-column deep neural networks for image classification. In: 2012 *IEEE Conference on Computer Vision and Pattern Recognition (CVPR)*, pp. 3642–3649, IEEE, (2012)
4. Deng, L. Chen, J.: Sequence classication using the high-level features extracted from deep neural networks. In: 2014 IEEE International Conference on Acoustics, Speech and Signal Processing (ICASSP), pp. 6844–6848, IEEE (2014)
5. Deng, L., Yu, D.: Deep learning: methods and applications. Found. Trends Signal Process. **7**(3–4), 197–387 (2013). ISSN: 1932–8346. doi:10.1561/2000000039
6. Hinton, G., Osindero, S., Teh, Y.-W.: A fast learning algorithm for deep belief nets. Neural Comput. **18**(7), 1527–1554 (2006)
7. Hinton, G., Deng, L., Yu, D., Dahl, G.E., Mohamed, A.-R., Jaitly, N., Senior, A., Vanhoucke, V., Nguyen, P., Sainath, T.N., et al.: Deep neural networks for acoustic modeling in speech recognition: the shared views of four research groups. Signal Process. Mag. IEEE **29**(6), 82–97 (2012)
8. Hubert, L., Arabie, P.: Comparing partitions. J. Classif. **2**(1), 193–218 (1985). ISSN: 0176–4268. doi:10.1007/BF01908075
9. Krizhevsky, A. Sutskever, I., Hinton, G.E.: Imagenet classication with deep convolutional neural networks. In: Advances in Neural Information Processing Systems, pp. 1097–1105 (2012)
10. Leonard, R.: A database for speaker-independent digit recognition, vol. 9, pp. 328–331. March 1984. doi:10.1109/ICASSP.1984.1172716
11. Mahendran, A., Vedaldi, A.: Understanding deep image representations by inverting them. In: CoRR (2014). arxiv:1412.0035
12. Rand, W.M.: Objective criteria for the evaluation of clustering methods. J. Am. Stat. Assoc. **66**(336), 846–850 (1971)
13. Tüske, Z., Golik, P., Schlüter, R., Ney, H.: Acoustic modeling with deep neural networks using raw time signal for LVCSR. In: Proceedings of the Annual Conference of International Speech Communication Association (INTERSPEECH) (2014)
14. Zeiler, M.D., Fergus, R.: Visualizing and understanding convolutional networks. In: Computer Vision-ECCV, pp. 818–833. Springer (2014)

ELM Based Algorithms for Acoustic Template Matching in Home Automation Scenarios: Advancements and Performance Analysis

**Giulio della Porta, Emanuele Principi, Giacomo Ferroni,
Stefano Squartini, Amir Hussain and Francesco Piazza**

Abstract Speech and sound recognition in home automation scenarios has been gaining an increasing interest in the last decade. One interesting approach addressed in the literature is based on the template matching paradigm, which is characterized by ease of implementation and independence on large datasets for system training. Moving from a recent contribution of some of the authors, where an Extreme Learning Machine algorithm was proposed and evaluated, a wider performance analysis in diverse operating conditions is provided here, together with some relevant improvements. These are allowed by the employment of supervector features as input, for the first time used with ELMs, up to the authors' knowledge. As already verified in other application contexts and with different learning systems, this ensures a more robust characterization of the speech segment to be classified, also in presence of mismatch between training and testing data. The accomplished computer simulations confirm the effectiveness of the approach, with F_1-Measure performance up to 99 % in the multicondition case, and a computational time reduction factor close to 4, with respect to the SVM counterpart.

1 Introduction

Nowadays, security, automation and easy-to-use human-machine interfaces (HMI) are the main keywords of the new home automation oriented technologies. In this context, efficient and reliable speech/sound recognition plays a crucial role, both for providing a user-friendly interface and for the detection of dangerous situations, such as persons' falls. One of the fundamental part of these technologies is provided by the

G. della Porta · E. Principi · G. Ferroni · S. Squartini (✉) · F. Piazza
Department of Information Engineering, Università Politecnica delle Marche,
Via Brecce Bianche, 60131 Ancona, Italy
e-mail: s.squartini@univpm.it

A. Hussain
Department of Computing Science and Mathematics, University of Stirling,
Stirling FK9 4LA, UK
e-mail: ahu@cs.stir.ac.uk

© Springer International Publishing Switzerland 2016
A. Esposito et al. (eds.), *Recent Advances in Nonlinear Speech Processing*,
Smart Innovation, Systems and Technologies 48,
DOI 10.1007/978-3-319-28109-4_16

159

automatic recognition of sentences uttered by the user as well as the identification of dangerous situations by solely exploiting the sound captured by one or more microphones [13].

In the last decades, many research efforts have been leading to a wide variety of speech recognition solutions able to recognise the human speech also in adverse conditions with a high reliability grade. The core of many of these systems is represented by Hidden Markov Models (HMM), while recently deep neural networks gained a significant attention [7, 17]. In addition to them, template matching techniques [16] have also been devoted a certain attention in the recent years [2, 11, 18]. In this paper, we focus on this approach due to its ease of implementation and independence on large dataset for system training. A simple template matching algorithm operates by measuring the distance between the input utterance and a set of templates with Dynamic Time Warping (DTW) [16]. A threshold is then used to discriminate between in-domain and out-of-domain sentences and to determine the input class affiliation. The original DTW technique presents some shortcomings such as the high computational cost and the low performance in speaker independent task. In recent works [2, 11, 18], however, efficient versions of the algorithm have been developed to overcome those negative aspects.

The template matching algorithm can be improved by opportunely employing discriminative techniques such as Support Vector Machine (SVM) [4, 5] or Extreme Learning Machine (ELM). However, these techniques cannot be directly employed since input utterances generally have different lengths, thus different number of feature vectors. The solutions to this issue are either based on hybrid SVM/HMM architectures [6] or on dynamic kernels [5]. Due to its capability to increase the training speed with respect to traditional neural networks learning methods, ELMs have recently gained much interest in the scientific community [9]. Moreover, a recent study [3] has shown that ELMs achieve performance similar to SVMs, but requiring reduced training and testing times.

This paper extends a previous work [14] by some of the authors in which acoustic template-matching approaches for automatic emergency detection were presented. In particular, in [14] DTW distances and the outerproduct of trajectory matrix have been employed as input to ELM and ELM-kernel classifiers. Here, the approach is improved by introducing Gaussian mean supervectors (GMS) [12] obtained by training a Gaussian Mixture Model (GMM) for representing the acoustic space and by collecting its mean values vector after adaptation with the Maximum a Posteriori algorithm. Moreover, the approach has been evaluated in different vocal efforts scenarios. The experiments have been conducted using the ITAAL [15] and APASCI [1] datasets. The former is an Italian speech corpus of home automation commands and distress calls recorded with distant and close-talking microphones containing speech signals uttered with normal and loud vocal efforts. The APASCI dataset is a larger Italian corpus used for creating the GMM Universal Background Model (UBM). Compared to [14], where the algorithms were tested in a *matched* acoustic scenario, here we extend the evaluation by introducing a *mismatched* and a *multicondition* scenario, thus performing a deeper performance analysis.

The paper outline is the following: the proposed approach description is provided in Sect. 2, whilst Sect. 3 illustrates the experiment setup, the dataset and the assessed results. Finally, Sect. 4 concludes the paper.

2 The Proposed Approach

The proposed approach is presented in this section following a rigorous formulation of the problem we aim to address. Let $\mathbf{U}_k = \{\mathbf{u}_{k,1}, \ldots, \mathbf{u}_{k,L_k}\}$ be an utterance composed of L_k low-level feature vectors $\mathbf{u}_{k,l}$ of dimension $D \times 1$ and l being the time frame index. A training $\mathcal{T}$ corpus can then be defined as:

$$\mathcal{T} = \{(\mathbf{U}_1, C_1), \ldots, (\mathbf{U}_K, C_K)\}, \tag{1}$$

where C_k is the class of utterance $\mathbf{U}_k$. Given a test utterance $\mathbf{Y} = \{\mathbf{y}_1, \mathbf{y}_2, \ldots, \mathbf{y}_{L_y}\}$, the problem is finding the corresponding label $C_y \in \{C_1, C_2, \ldots, C_K\}$ based on a certain classification criterion.

In this paper, two different classifiers have being employed: ELM and ELM with kernel. Generally, each input utterance is composed of a different number of feature vectors L_k, thus preventing the use of the aforementioned classifiers without using a length normalisation algorithm. In particular, each utterance is here mapped to a fixed-length feature vector by employing the Gaussian Mean supervector of the utterance. As low-level feature set, the MFCC is employed, largely used in speech recognition tasks (Fig. 1).

Fig. 1 Block scheme of the proposed approach

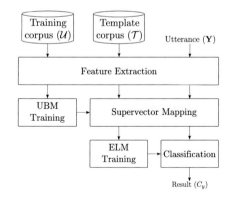

2.1   Gaussian Mean Supervectors

Gaussian Mean Supervectors are an effective and compact way for representing an input utterance. In order to calculate the GMS of input utterance, a Universal Background Model (UBM) representing a statistical description of the acoustic space is needed. The Gaussian Mixture Model (GMM) representing the UBM is given by the following expression:

$$p(\mathbf{x}|\lambda) = \sum_{j=1}^{J} w_j p(\mathbf{x}|\boldsymbol{\mu}_j, \boldsymbol{\Sigma}_j), \tag{2}$$

where $\lambda = \{w_j, \boldsymbol{\mu}_j, \boldsymbol{\Sigma}_j | j = 1, 2, \ldots, J\}$, w_j are the mixture weights, and $p(\cdot|\boldsymbol{\mu}_j, \boldsymbol{\Sigma}_j)$ is a multivariate Gaussian distribution with mean vector $\boldsymbol{\mu}_j$ of size $D \times 1$ and diagonal covariance matrix $\boldsymbol{\Sigma}_j$ of size $D \times D$.

The GMS $\mathbf{M}$ of an utterance $\mathbf{X} = \{\mathbf{x}_1, \mathbf{x}_2 \ldots, \mathbf{x}_L\}$ composed of L low-level feature vectors (e.g., MFCCs) is obtained by adapting the means of the UBM model with maximum a posteriori (MAP) adaptation and then concatenating the mean vectors:

$$\mathbf{M} = [\boldsymbol{\mu}_1^T, \boldsymbol{\mu}_2^T, \cdots, \boldsymbol{\mu}_J^T]^T,$$

where T denotes the transpose operator. Regardless the length of the input utterance, $\mathbf{M}$ is a $DJ \times 1$ vector.

2.2   Extreme Learning Machine

ELM is a fast learning algorithm designed for single hidden layer feedforward neural networks (SLFNs). In ELM, the input weights of SLFNs do not need to be tuned. They can be randomly generated, whereas the output weights are analytically determined using the least-square method. This process allows a significant training time reduction.

Consider a set of N labelled training samples $\{(x_1, t_1), \ldots, (x_N, t_N)\}$ where $x_i \in \{-1, 1\}$, and SLFN with I input neurons and L hidden neurons (Fig. 2). The ELM decision function, for binary classification, is the following:

$$f_L(x) = \text{sign}\left(\sum_{i=1}^{L} \beta_i h_i(x)\right) = \text{sign}\left(h(x)\beta\right).$$

In the equation, the vector $\beta = [\beta_1, \ldots, \beta_L]^T$ contains the weight hidden neurons and output neurons, while $h(x) = [h_1(x), \ldots, h_L(x)]$ is the output of the hidden layer with respect to the input x. Usually, $h(x) = [G(a_1, b_1, x), \ldots, G(a_L, b_L, x)]$ and $G(a, b, x)$ is a nonlinear piecewise continuous function that satisfies ELM universal

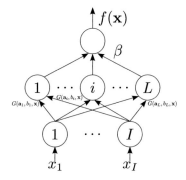

Fig. 2 ELM with I input neurons and L hidden neurons

approximation capability theorems, and $\{a_i, b_i\}_{i=1}^{L}$ are randomly generated. In order to provided the best performance in the experiments, $G(a_1, b_1, x)$ assumes the form of the sigmoid.

Defining the hidden-layer output matrix **H** as

$$\mathbf{H} = \begin{pmatrix} h_1(x_1) & \cdots & h_L(x_1) \\ \vdots & \vdots & \vdots \\ h_1(x_N) & \cdots & h_L(x_N) \end{pmatrix},$$

training th ELM consist in minimizing $||\mathbf{H}\beta - \mathbf{T}||$ and $||\beta$ where $\mathbf{T} = [t_1, t_2, ..., t_N]^T$. The solution to the problem can be calculated as the minimum norm least-square solution of the linear system:

$$\hat{\beta} = \mathbf{H}^{-1}\mathbf{T},$$

where $\mathbf{H}^{-1}$ is the Moore-Penrose generalized inverse of matrix **H**. ELM allows achieving good generalization performance, by computing output weights analytically, with speedy training phase.

2.2.1 Extreme Learning Machine with Kernels

In kernel-based ELM [8], $\mathbf{h}(\mathbf{x})$ is unknown, and the output function of the classifier is written as:

$$\mathbf{f}(\mathbf{x}) = \begin{bmatrix} K(\mathbf{x}, \mathbf{x}_1) \\ \vdots \\ K(\mathbf{x}, \mathbf{x}_N) \end{bmatrix}^T \left(\frac{\mathbf{I}}{C} + \mathbf{\Omega} \right)^{-1} \mathbf{T}, \tag{3}$$

where $\mathbf{\Omega}$ is defined so that each element $\Omega_{i,j} = h(\mathbf{x}_i) \cdot h(\mathbf{x}_j) = K(\mathbf{x}_i, \mathbf{x}_j)$. $K(\cdot, \cdot)$ is a kernel function as in SVM, and in this work assumes the form a radial basis function. It is interesting to note that differently from standard ELM, the number of hidden neurons must not be known in advance.

3 Experiments

The proposed approach has been evaluated on the ITAAL corpus [15]. The dataset is composed of utterances spoken by 20 native Italian speakers coming from the central Italy (Marche region). Recordings have been performed on a room with a reverberation time equal to 0.72 s using a headset microphone and an array composed of four microphones. Every sentence was spoken both with normal and loud vocal efforts. In the experiments, only the headset microphone has been used. The UBM has been trained on APASCI [1], an Italian speech database composed of 5,290 phonetically rich sentences and 10,800 isolated digits, for a total of 641 min of speech. The speech material was read by 100 Italian speakers and recorded in quiet conditions.

The experiment consists in evaluating the algorithms performance in three different tasks, *matched*, *mismatched* and *multicondition*, that differ in the composition of the training and test sets. In the *matched* task, both test and training sets are composed of signals uttered with the same vocal effort. On the contrary, in the *mismatched* task the training set and the test sets contain signals uttered with different vocal effort. In the remaining task, *multicondition*, testing and training signals comprise both vocal efforts. Regardless the task, the evaluation has been performed using a leave-one-speaker-out method, i.e., each speaker at turn is employed for testing and not for training.

The GMS-based approach has been compared to alternative techniques which employ a different length normalisation method and a different classifier. In particular, as in [14], DTW distances have been employed instead of supervectors and SVM is used as alternative classifier. The DTW length normalisation provides a fixed-length vector from the input utterance by calculating the distance between itself and the other utterances of the input set $\mathcal{T}$. Recalling the notation in Sect. 2, an input utterance $\mathbf{X}$ is mapped to an $I \times 1$ vector $\mathbf{v} = [d(\mathbf{X}, \mathbf{U}_1), d(\mathbf{X}, \mathbf{U}_2), \ldots, d(\mathbf{X}, \mathbf{U}_I)]^T$, where $\mathbf{U}_1 \ldots \mathbf{U}_I \in \mathcal{T}$ and $d(\cdot, \cdot)$ represents the DTW distance. Referring to Fig. 1, DTW-based length normalisation replaces the supervector mapping block. Clearly, this method does not require the UBM.

Regarding the parameters of feature extraction pipelines, the sample rate of the signals is 16 kHz and a pre-emphasis coefficient $\mu = 0.97$, a frame length of 25 ms and a hop size of 10 ms, and 40 filters in the mel-like filterbank are used.

The algorithms performance have been evaluated by varying the classifier-related parameters. In particular, the size of the ELM has been varied from 30 to 1000 neurons by gradually incrementing it of 20 neurons at each iteration. The optimal values of the ELM kernel and SVM parameters C and γ have been selected using a grid search. More specifically, both of them have been varied from 2^{-15} to 2^{15} incrementing the exponent by 2 at each iteration. Regarding the supervector extraction, the number of Gaussians of the UBM has been varied from 4 to 64 by doubling it at each iteration. Table 1 reports the parameter values for all classifiers in the diverse operating conditions addressed in our experiments. The LIBSVM library has been used to implement the SVM classifier, whilst an ANSI C implementation of the ELM and ELM kernel is used for ELM experiments. The performance has been assessed using

Table 1 Best parameter values for all classifier in performed simulations

	Matched		Mismatched		Multicondition
	Normal	Loud	Normal	Loud	
Distance-based					
ELM (neurons)	70	120	150	90	80
ELM kernel (C, γ)	$2^{-3}, 2^{-7}$	$2^7, 2$	$2^2, 2^{-1}$	$2^{-1}, 2^{-1}$	$2^2, 2^{-1}$
SVM (C, γ)	$2^7, 2^{-3}$	$2^5, 2^{-3}$	$2^{12}, 2^{-9}$	$2^5, 2^{-1}$	$2^6, 2^{-1}$
Supervector-based					
ELM (neurons, No. gaussians)	$280, 2^3$	$400, 2^3$	$480, 2^3$	$320, 2^3$	$620, 2^3$
ELM kernel $(C, \gamma,$ No. gaussians)	$2^{-1}, 2^5, 2^5$	$2^6, 2^5, 2^5$	$2, 2^{-8}, 2^5$	$2, 2^3, s^5$	$2, 2^3, s^5$
SVM $(C, \gamma,$ No. gaussians)	$2, 2^{-6}, 2^5$	$2, 2^{-6}, 2^5$	$2, 2^{-8}, 2^5$	$2, 2^{-6}, 2^4$	$2, 2^{-7}, 2^5$

the average of F_1-Measure over the number of classes (i.e., the type of sentences) in the dataset.

3.1 Results and Discussion

In this subsection the results obtained in performed computer simulations are reported.

Table 2 shows the obtained results for the *matched* case. Part of these results have been already presented in [14], and they are now completed with those related to supervectors-based feature extraction, for all involved classifiers. It can be easily observed that ELM kernel and SVM present very close performance, both in "distance" and "supervectors" case studies, always outperforming ELM except for the "distance-Loud" operating condition.

Table 3 shows the results obtained on the *mismatched* task. With the exception of ELM kernel trained on normal signals and tested on loud signals, all the algorithms present a performance decrease respect to the matched condition. Interestingly, the

Table 2 F_1-Measure (%) obtained on the *matched* task

	Distance		Supervectors	
	Normal	Loud	Normal	Loud
ELM	88.45	90.45	81.38	85.06
ELM kernel	90.97	86.32	98.40	98.56
SVM	91.78	87.16	98.69	99.02

Table 3 F_1-Measure (%) on the *mismatched* task

	Distance		Supervectors	
	Normal	Loud	Normal	Loud
ELM	83.24 (−5.21)	89.06 (−1.39)	72.49 (−8.89)	73.68 (−11.38)
ELM kernel	89.31 (−1.66)	91.07 (+4.75)	92.56 (−5.84)	91.89 (−6.67)
SVM	81.11 (−10.67)	85.46 (−1.70)	95.16 (−3.53)	95.54 (−3.48)

The "Normal" (respectively, "Loud") column results have obtained training the algorithms on "Loud" (respectively, "Normal") signals. The difference between the matched results of Table 2 is shown in brackets

three classifiers exhibit a different behaviour, with SVM suffering the most from the signal mismatch with an average F_1-Measure decrease amounting to 6.20 %. The ELM decrease is about the half, 3.30 %, while ELM kernel improves the performance by 1.55 % thanks to the increase in the loud signals test.

Table 4 shows the results for the multicondition task. In this case ELM kernel offers remarkable results: indeed it allows to achieve, in the "supervector" case study, an accuracy higher than 7 % with respect to the "distance" case and even slightly better than the ones obtained in the *matched* condition (see Table 2). Moreover, whereas ELM kernel and SVM show comparable performance (F_1-Measure up to 99 %), ELM does not seem to compete with them.

As pointed out in the literature, one of the advantages of using ELM with respect to SVM relies on the execution times, so it is worth analysing the performance of the algorithms from this perspective. The results shown in Table 5 represent the time required to perform both training and testing, in the *matched* condition with the best parameter values for all involved classifiers (see Table 1). It is evident from the values in Table 5 that ELM and ELM Kernel are the most performing algorithms. In particular, with the usage of supervectors, ELM Kernel allows to reduce the computational time by a factor of 4. Similar conclusions can be drawn in the *mismatched* and *multicondition* conditions.

Table 4 F_1-Measure (%) on the *multicondition* task

	Distance	Supervectors
ELM	89.20	87.68
ELM kernel	91.78	99.06
SVM	91.82	99.05

Table 5 Execution times (s)

	Distance	Supervectors
ELM	0.018	0.248
ELM kernel	0.022	0.120
SVM	0.032	0.452

4 Conclusion

In this paper, the authors extend the approach recently proposed in [14] for acoustic template matching in home automation scenarios, by involving the supervector paradigm for feature length normalization in ELM training. Up to the authors' knowledge, this is the first attempt in the literature along this direction, and can thus represent a valuable reference also for diverse application fields.

ELM and ELM kernel algorithms have been compared with the standard SVM, by using MFCCs, the popular choice in many automatic speech recognition systems, as low level features, and DTW distance and supervectors as feature length normalization techniques. The experiments to assess the performance of the algorithms have been conducted against the ITAAL corpus. The APASCI database has been used for creating the UBM, needed for the calculation of supervectors.

Differently from [14] and exploiting the two distinct vocal efforts (normal vs. loud) included in the ITAAL database, three different operating conditions have been addressed in this work: the matched, the mismatched and the multicondition. The results demonstrate the good performance, both in terms of accuracy (F_1-Measure up to 99 % in the multicondition case) and execution times (computational time reduction factor close to 4, with respect to SVM), of the ELM kernel algorithm by using the supervectors approach for feature length normalization.

In future works, further experiments will be carried out by involving alternative feature sets (i.e. Power Normalized Cepstral Coefficients) and also employing different learning systems, like Deep Neural Networks [7] and Echo State Networks [10].

References

1. Angelini, B., Brugnara, F., Falavigna, D., Giuliani, D., Gretter, R., Omologo, M.: Automatic segmentation and labeling of english and italian speech databases. In: Proceedings of Eurospeech, pp. 653–656. Berlin, Germany, 22–25 Sept 1993
2. Anguera, X.: Information retrieval-based dynamic time warping. In: Proceedings of Interspeech, pp. 1–5. Lyon, France, 25–29 Aug 2013
3. Chorowski, J., Wang, J., Zurada, J.M.: Review and performance comparison of SVM-and ELM-based classifiers. Neurocomputing **128**, 507–516 (2014)
4. Cortes, C., Vapnik, V.: Support-vector networks. Mach. Learn. **20**(3), 273–297 (1995)
5. Dileep, A.D., Sekhar, C.C.: Class-specific GMM based intermediate matching kernel for classification of varying length patterns of long duration speech using support vector machines. Speech Commun. **57**, 126–143 (2014)
6. Ganapathiraju, A., Hamaker, J., Picone, J.: Hybrid SVM/HMM architectures for speech recognition. In: Proceedings of ICSLP, pp. 504–507. Beijing, China, 16–20 Oct 2000
7. Hinton, G., Deng, L., Yu, D., Dahl, G.E., Mohamed, A.r., Jaitly, N., Senior, A., Vanhoucke, V., Nguyen, P., Sainath, T.N., et al.: Deep neural networks for acoustic modeling in speech recognition: the shared views of four research groups. Signal Process. Mag., IEEE **29**(6), 82–97 (2012)
8. Huang, G.B., Zhou, H., Ding, X., Zhang, R.: Extreme learning machine for regression and multiclass classification. IEEE Trans. Syst., Man, Cybern. B **42**(2), 513–529 (2012)

9. Huang, G.B., Zhu, Q.Y., Siew, C.K.: Extreme learning machine: theory and applications. Neurocomputing **70**(1), 489–501 (2006)
10. Jaeger, H.: The "echo state" approach to analysing and training recurrent neural networks. Tech. Rep. 148, German National Research Center for Information Technology, Bonn, Germany (2001)
11. Kim, C., Seo, K.D.: Robust DTW-based recognition algorithm for hand-held consumer devices. IEEE Trans. Consum. Electron. **51**(2), 699–709 (2005)
12. Kinnunen, T., Li, H.: An overview of text-independent speaker recognition: from features to supervectors. Speech Commun. **52**(1), 12–40 (2010)
13. Principi, E., Squartini, S., Bonfigli, R., Ferroni, G., Piazza, F.: An integrated system for voice command recognition and emergency detection based on audio signals. Expert Syst. Appl. **42**(13), 5668–5683 (2015)
14. Principi, E., Squartini, S., Cambria, E., Piazza, F.: Acoustic template-matching for automatic emergency state detection: an ELM based algorithm. Neurocomputing **149**, 426–434 (2014)
15. Principi, E., Squartini, S., Piazza, F., Fuselli, D., Bonifazi, M.: A distributed system for recognizing home automation commands and distress calls in the Italian language. In: Proceedings of Interspeech, pp. 2049–2053. Lyon, France, 25–29 Aug 2013
16. Rabiner, L., Juang, B.H.: Fundamentals of Speech Recognition. Prentice Hall PTR (1993)
17. Saon, G., Chien, J.T.: Large-vocabulary continuous speech recognition systems: a look at some recent advances. IEEE Signal Process. Mag. **29**(6), 18–33 (2012)
18. Zhang, X., Sun, J., Luo, Z., Li, M.: Confidence Index Dynamic Time Warping for Language-Independent Embedded Speech Recognition. In: Proceedings of ICASSP, pp. 8066–8070. Vancouver, Canada, 26–31 May 2013

Linear Versus Nonlinear Multi-scale Decomposition for Co-channel Speaker Identification System

Wajdi Ghezaiel, Amel Ben Slimane and Ezzedine Ben Braiek

Abstract Co-channel speech is a combination of speech utterances over a single communication channel. Traditional approach to co-channel speech processing is to attempt to extract the speech of the speaker of interest (target speech) from other (interfering) speech. Usable speech criteria are proposed to extract minimally corrupted speech for speaker identification in co-channel speech. In this paper, we present usable speech extraction method based on pitch information obtained from linear multi-scale decomposition by dyadic wavelet transform and nonlinear multi-scale decomposition by empirical mode decomposition. Detected usable speech are organized into speaker stream, and applied to speaker identification system. The proposed methods are evaluated and compared across various Target to Interferer Ratio (TIR) for speaker identification system.

1 Introduction

Degrading the quality and intelligibility of the speech signals, background noise is a severe problem in communication and related speech systems. The desired signal is mostly contaminated with some interference sources. There are different types of noise signals which affect the quality of the original speech. Noise signals can be classified as stationary or non stationary. Stationary noise can be dealt with by using denoising and noise reduction techniques; whereas non stationary noise is caused by another speech from a different speaker. Such interference is frequent and the corrupted speech is known as co-channel speech [1]. In traditional speaker identification

W. Ghezaiel (✉) · E. Ben Braiek
CEREP, ENSIT University of Tunis, Tunis, Tunisia
e-mail: wajdi.ghezaiel@gmail.com

E. Ben Braiek
e-mail: Ezzedine.Benbraiek@esstt.rnu.tn

A. Ben Slimane
ENSI University of Mannouba, Mannouba, Tunisia
e-mail: Amel.benslimane@ensi.rnu.tn

© Springer International Publishing Switzerland 2016
A. Esposito et al. (eds.), *Recent Advances in Nonlinear Speech Processing*,
Smart Innovation, Systems and Technologies 48,
DOI 10.1007/978-3-319-28109-4_17

SID, only one target speaker exists in the given signal whereas in co-channel SID, the task is to identify the target speakers in one given mixture. Research on co-channel speaker identification has been done for more than one decade [1], yet the problem remains largely unsolved. Research has been carried to extract one of the speakers from co-channel speech by either enhancing target speech or suppressing interfering speech [1]. In automatic speaker recognition, as pointed out in [2], the intelligibility and quality of extracted speech are not important. What the system needs are portions of the speech that contain speaker characteristics unique to an individual speaker, classifiable and long enough for the system to make identification or verification decisions. These portions of speech, or segments, are defined as consecutive frames of speech that are minimally corrupted by interfering speech and are, thus, called usable speech [2]. Yantormo performed a study on co-channel speech and concluded that the target to interferer ratio TIR was a good measure to quantify the usability for speaker identification [3]. The TIR is a power ratio of the target speech to the interfering speech. This ratio can be expressed for entire utterances or individual frames of speech. For usability, frames above 20 dB TIR are considered usable. However the TIR is not an observable value from the co-channel speech data. Hence, a number of methods for usable speech detection which refer to the TIR have been developed and studied under co-channel condition [4–6]. In these methods, usable speech frames are composed of voiced speech. In [7], the Peak difference autocorrelation of wavelet transform method (PDAWT) is applied in order to detect pitch information in usable speech. This method applies autocorrelation on approximation component obtained by filtering co-channel speech at one dyadic wavelet transform (DWT) scale. In our previous work [8, 9], we have developed linear multi-scale decomposition by dyadic wavelet transform (MRDWT) method to detect usable speech. MRDWT method applies dyadic wavelet transform (DWT) iteratively to detect pitch periodicity. We are motivated by detecting pitch information in all lower frequency sub-bands of co-channel speech. DWT is performed for stationary signal analysis. However co-channel speech signal is non stationary signal, a nonlinear multi-scale approach which incorporates the Empirical Mode Decomposition (EMD) may be effective [10]. EMD is a signal processing technique particularly suitable for non-linear and non-stationary signal, has recently been proposed [11] as a new tool for data analysis. The EMD method is able to decompose a complex signal into a series of intrinsic mode functions (IMF) and a residue in accordance with different frequency bands [11]. We have proposed in [12–14] a new method for usable detection by empirical mode decomposition MREMD. In this paper, we propose to compare performance of the two proposed usable speech method for speaker identification system. Evaluation of this method is performed on TIMIT database referring to the TIR measure. Co-channel speech is constructed by mixing all possible gender speakers. Discussion of proposed methods are provided basing on evaluation results. The next section we describe how to extract usable speech using linear and nonlinear multi-scale decomposition. In Sect. 3, we present the speaker identification system. Experiment results and comparisons are given in Sect. 4. Section 5 concludes the paper.

2 Multi-scale Decomposition for Usable Speech Detection

Usable frames are characterized by periodicity features. These features should be located in low-frequency band that includes the pitch frequency. Multi-scale decomposition is applied iteratively in order to determine the suitable band for periodicity detection. In this low-frequency band, periodicity features are not much disturbed by interferer speech in case of usable segments. In case of unusable frames, it is not possible to detect periodicity in all lower sub-bands. At each iteration, autocorrelation is applied to the low-frequency band in order to detect periodicity [15]. Three dominated local maxima are determined from the autocorrelation signal with a peak-picking algorithm which uses a threshold calculated from local maxima amplitudes. A difference of autocorrelation lag between the first and second maximum and between the second and third maximum is determined. If this difference is less than the threshold, periodicity is detected and co-channel speech segment is classified as usable. This threshold is empirically fixed according to the best evaluation results. The optimum threshold value of 8 samples is chosen at 16 kHz sampling frequency. If at this scale, periodicity is not detected, a multi-scale decomposition is applied to this low-frequency band signal in order to detect hidden periodicity feature in finer band frequency. For unusable frames, it is not possible to detect periodicity in all lower sub-bands.

2.1 Linear Multi-scale Decomposition by Dyadic Wavelet

Linear Multi-scale decomposition based on dyadic wavelet transform (MRDWT) is used to decompose voiced co-channel speech into a linear combination of two components [8, 9]. The first component ranging from the high-frequency band and called detail. The second component ranging to low-frequency band and called approximation. Autocorrelation is applied on approximation to detect pitch information [15]. A maximum of 4 iterations are allowed. This limit is fixed based on pitch band. The lowest band should correspond to pitch band. Figure 1 shows usable frame for male-male co-channel speech. Figure 1 shows that periodicity is not detected respectively at scale 1 and scale 2. It is noted that the fundamental periodicity of the voiced speech becomes clearer in the correlation domain. In this case, periodicity is detected only at scale 3. Hence this frame is classified as usable.

2.2 Nonlinear Multi-scale Decomposition by Empirical Mode Decomposition

Nonlinear Multi-scale decomposition based on EMD (MREMD) is used to decompose voiced co-channel speech into a linear combination of two components. The first

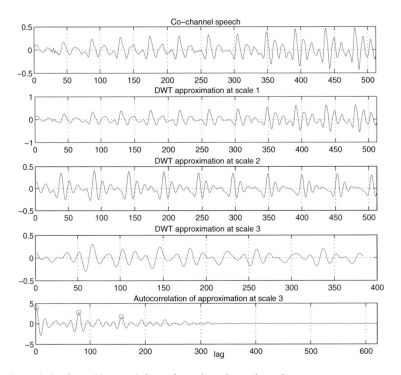

Fig. 1 Analysis of a usable speech frame for male-male co-channel

component called intrinsic mode function (IMF) ranging from the high-frequency band and so-called detail [12]. The second component called residue ranging to low-frequency band and so-called approximation [12]. Autocorrelation is applied on approximation to detect pitch information [13, 14]. A maximum of 5 iterations are allowed. This limit is fixed empirically basing on evaluation results, the lowest band should correspond to pitch band. Figure 2 shows usable frame for female-male co-channel speech. Periodicity is not detected respectively at scale 1 and scale 2. Periodicity is detected only at scale 3.

3 Speaker Identification System

In order to identify the target and the interferer speakers, the detected usable segments are organized into two speaker streams by a speaker assignment system [16]. The speaker assignment system organizes usable speech segments under co-channel conditions. It has extended probabilistic framework of traditional SID to co-channel speech. It uses exhaustive search algorithm to maximize the posterior probability in

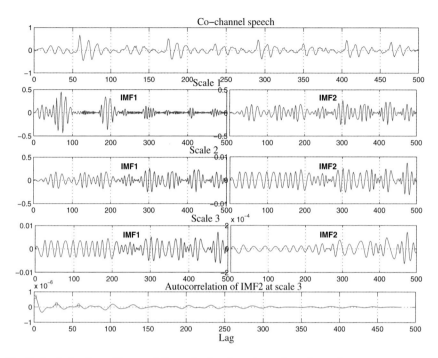

Fig. 2 Analysis of a usable speech frame for female-male co-channel speech

grouping usable speech. Then, usable segments are assigned to two speaker groups, corresponding to the two speakers in the mixture. The two speaker streams are used as input for speaker identification system. The SID is performed with a baseline system [17]. Modeling is assured by Gaussian Mixture Model (GMM) and estimated through the Expectation Maximization (EM) algorithm that maximizes the likelihood criterion. A set of 16 mixtures are used for speaker model. In our experiment, we use the classical parameterization based on 16 Mel Frequency Cepstral Coefficients (MFCC). These coefficients are computed from the speech signal every 10 ms using a time window of 25 ms. Each feature vector is presented by the middle windows of every utterance. Speaker model is trained using the EM algorithm with the features calculated from training samples. In testing phase, the organized usable speech, with speaker assignment system, are used as test speech samples for SID system. The same features are derived from the test speech samples and are input to every speakers GMM. The speaker with the highest likelihood score represents the identified speaker. Here, speaker identification experiments are close-set and text-independent.

4   Evaluation Results

The TIMIT database is used for all the simulation experiments. The TIMIT database is just used for illustration purposes like in [2]. The speaker set is composed of 38 speakers from the DR1? dialect region, 14 of which are female and the rest are male. Each speaker has 10 utterance files, 5 out of 10 files are used for training and the remaining 5 files are used to create co-channel mixtures for testing. For each speaker deemed as the target speaker, 1 out of 5 test files is randomly selected and mixed with randomly selected files of every other speaker, which are regarded as interfering utterances. For each pair, the TIR is calculated as the energy ratio of the target speech over the interference speech. Three different sets of co-channel speech are considered: male-male, female-female, and male-female. Thus, for each TIR, a total of 1406 co-channel mixture files are created for the testing purpose.

4.1   Evaluation of Usable Speech Detection Methods

The Target to Interferer Ratio TIR measure is used to label voiced frames as usable or unusable. For usability decision, frames that have above 20 dB TIR are considered as usable. Evaluation is based on hits and false alarms percentages. The performance of the proposed methods is given in Table 1.

Peak difference autocorrelation of wavelet transform (PDAWT) method [7] applies DWT once only to co-channel speech to detect pitch information. On average the PDAWT method detects at least 81 % of the usable speech with a false alarm rate of 30 %. On average the MRDWT method detects at least 95.76 % of the usable speech with a false alarm rate of 29.65 % [7]. Nonlinear multi-scale decomposition by EMD [12–14] achieves a minimum of false alarm compared to MRDWT and PDAWT methods. We consider the effectiveness of EMD to reduce the percent of false alarm. EMD achieve a maximum of hits for usable speech detected in male male co-channel. We show the effectiveness of the nonlinear multi-scale decomposition to detect usable speech.

Table 1   Results of usable speech detection methods

Co-channel speech	PDAWT		MRDWT		MREMD	
	% hits	% FA	% hits	% FA	% hits	% FA
Female-female	82.00	32.30	93.02	32.37	98.20	15.34
Male-male	80.50	30.60	98.46	28.93	98.69	18.60
Male-female	81.30	29.60	95.80	27.66	99.03	13.62
Average	81.20	30.80	95.76	29.65	98.64	15.85

Fig. 3 Performance of the
proposed speaker
identification under
co-channel conditions
compared with related
methods

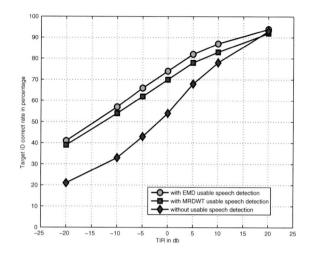

4.2 Speaker Identification Evaluation

The SID identifies the target and the interferer speaker. The evaluation system aims
to evaluate the identification of the speaker referred as target.

It clear from Fig. 3 that the EMD performs significantly better than the MRDWT
usable speech method. This effectiveness is due to reduction of false alarm by our
proposed method. The target SID correct rate with usable speech detection is better than the target SID correct rate without usable speech detection. The proposed
usable speech detection improves speaker identification performance. The average
improvement is about 16 % in terms of SID correct rate. Also the improvements are
consistent across all TIR levels. Performance improvement increases at higher TIR
because the target speaker dominates the mixture. However, target speaker is dominated by interference at negative TIR, resulting in better performance after usable
speech extraction. The accuracy degrades sharply when TIR decreases because the
target speech becomes increasingly corrupted.

5 Conclusion

In this paper, we have proposed a speaker identification system in co-channel speech.
We have proposed a new usable speech detection method based on multi-scale decomposition by linear dyadic wavelet transform MRDWT and nonlinear empirical mode
decomposition MREMD. Usable speech is extracted based on the pitch information obtained from sub-band analysis. Our usable speech extraction methods produces segments useful for co-channel SID across various TIR conditions. MREMD
achieves a good percent of usable speech detection and a minimum of false alarms.

In comparison with wavelet linear filtering methods MRDWT, MREMD method achieves a minimum of false alarms. We note the effectiveness of a nonlinear multi-scale decomposition by EMD to reduce the percentage of false alarms. Usable segments are assigned to two speaker groups, corresponding to the two speakers in the mixture. Organized usable speech are used as input to speaker identification system. We have shown that proposed usable speech detection methods achieves good SID performance and it performs significantly better than without usable speech detection. SID performance degrades when TIR decreases because the target speech is increasingly corrupted by interferer speech. We consider the effectiveness of MREMD to achieve good SID performance and it performs significantly better than MRDWT method for usable speech detection.

References

1. Quatieri, Danisewicz: An approach to co-channel talker interference suppression using a sinusoidal model for speech. IEEE Trans. Acoust. Speech Sig. Process. **38**, 56–69 (1990)
2. Lovekin, J., Yantorno, R.E., Benincasa, S., Wenndt, S., Huggins, M.: Developing usable speech criteria for speaker identification. Proc. ICASSP 421–424 (2001)
3. Yantorno, R.E.: Method for improving speaker identification by determining usable speech. J. Acoust. Soc. Am. 124 (2008)
4. Krishnamachari, K.R., Yantorno, R.E., Benincasa, D.S., Wenndt, S.J.: Spectral autocorrelation ratio as a usability measure of speech segments under cochannel conditions. IEEE Int. Symp. Intell. Sig. Process. Commun. Syst. (2000)
5. Lovekin, J., Krishnamachari, K.R., Yantorno, R.E., Benincasa, D.S., Wenndt, S.J.: Adjacent pitch period comparison (APPC) as a usability measure of speech segments under cochannel conditions. IEEE Intell. Sig. Process. Commun. Syst. 139–142 (2001)
6. Smolenski, B.Y., Ramachandran, R.P.: Usable speech processing: a filterless approach in the presence of interference. IEEE Circuits Syst. Mag. (2011)
7. Kizhanatham, Yantorno, R.E.: Peak difference autocorrelation of wavelet transform algorithm based usable speech measure, 7th world multi-conference on systemic, cybernetics, and informatics, (2003)
8. Ghezaiel, W., Ben Slimane, A., Ben Braiek, E.: Usable speech detection for speaker identification system under co-channel conditions, international conference on electrical system and automatic control JTEA, Tunisia (2010)
9. Ghezaiel, W., Ben Slimane, A., Ben Braiek, E.: Evaluation of a multi-resolution dyadic wavelet transform method for usable speech detection. World Acad. Sci. Eng. Technol. J. WASET 829–833 (2011). pISSN:2010-376X, eISSN:2010-3778
10. Huang, N.E., Shen, Z., Long, S.R., et al.: The empirical mode decomposition and Hilbert spectrum for nonlinear and non-stationary time series analysis. Proc. R. Soc. Lond. A **454**, 903–995 (1998)
11. Flandrin, P., Rilling, G., Goncalves, P.: Empirical mode decomposition as a filter bank. IEEE Sig. Process. Lett. **11**(2), 112114 (2004)
12. Ghezaiel, W., Ben Slimane, A., Ben Braiek, E.: Usable speech detection based on empirical mode decomposition. IET Electron. Lett. **49**(7), (2013)
13. Ghezaiel, W., Ben Slimane, A., Ben Braiek, E.: Multi-resolution analysis by empirical mode decomposition for usable speech detection, international multi-conference on systems, signals devices, conference on communication signal processing, SSD, Tunisia (2013)
14. Ghezaiel, W., Ben Slimane, A., Ben Braiek, E.: Improved, E.M.D., usable speech detection for co-channel speaker identification. Lecture Notes in Computer Science, Advances in Non-

Linear Speech Processing, 7911, pp. 184–191; International Conference: Non Linear Speech Processing 2013, (NOLISP 2013). Mons, Belgium (2013)

15. Hess, W.H.: Pitch Determination of Speech Signal: Algorithms and Devices. Springer, Heidelberg (1983)
16. Ghezaiel, W., Ben Slimane, A., Ben Braiek, E.: Usable speech assignment for speaker identification under co-channel situation. Int. J. Comput. Appl. **59**(18), 7–11, (2012)
17. Reynolds, D.A.: Speaker identification and verification using Gaussian mixture speaker models. Speech Commun. **17**, 91108 (1995)

Part V
Identifying New Nonlinear Coding and Decoding Features

Wigner-Ville Representation of a Stationary Voiced Speech Model

Khawla Zammel and Noureddine Ellouze

Abstract The Wigner Ville transform is introduced as an appropriate tool to analyze the signal in time and frequency plan since it give an accurate representation of energy. This article aims at providing a new mathematical representation for voiced speech signal based on Wigner Ville transform (WVT). We will create in the first place a mathematical formulation of the Wigner Ville Transform of a stationary voiced speech model. The speech model is expressed as a linear combination of sine waves with amplitudes, phases and harmonic frequencies. We then will compare the WVT of a real voiced speech signal to the WVT terms of the previous model.

Keywords Wigner Ville Transform · Voiced speech model · Fourier transform · Energy distribution

1 Introduction

The interest of time and frequency representation of a signal has been widely acknowledged. It concerns the analysis and processing of signals with time-varying frequency content. The purpose of this representation is to give information about how the energy of the signal is distributed in the time-frequency space [1]. Analyzing speech signal is a matter of particular interest dues to the amount of information that the signal hold, but all the difference is made by choosing the right way to proceed with the analysis.

Fourier and the spectrogram (squared modulus of the Fourier transform) are the most intuitive approaches for time-frequency analysis. However, several other signal

K. Zammel (✉) · N. Ellouze (✉)
Signal, Image and Pattern Recognition Lab, National School of Engineers
of Tunis ENIT, Tunis, Tunisia
e-mail: khawla.zammel.2014@ieee.org
URL: http://www.enit.rnu.tn/

N. Ellouze
e-mail: n.ellouze@enit.rnu.tn

© Springer International Publishing Switzerland 2016
A. Esposito et al. (eds.), *Recent Advances in Nonlinear Speech Processing*,
Smart Innovation, Systems and Technologies 48,
DOI 10.1007/978-3-319-28109-4_18

process methods were developed [7]. In the case of the quadratic distributions, the various admissible solutions are grouped in a set called Cohen class. These distributions are covariant in time and frequency. In order to proceed with a time-frequency representation, it is needed to choose between interferences (bilinear nature of the distribution) and spreading (after smoothing to lessen the interferences) and each analysis method has its advantages and disadvantages.

The Wigner Ville Transform (WVT) is a good candidate for time-frequency analysis. It was introduced by Ville in 1948 in signal processing in order to improve time-frequency concentration and was first considered by Wigner in 1932 in the field of quantum thermodynamics [12].

In this paper, we will cover first the modelling of the voiced speech signal model based on its Fourier transform. We then will compare the WVT transform of a real voiced speech signal versus the WVT of the model created. We will highlight a large number of WVT properties and all the questions that arises. Our challenge, through creating such a model, is to explore from a theoretical point of view the power of Wigner Ville applied on speech signal as a time frequency representation and some of its properties.

2 Time-Frequency Analysis of a Voiced Model Speech Signal

The idea of modelling the speech signal initiated by Portnoff [16] is also the concern of many research made in the 1980s; including, mainly, those by McAulay and Quatieri [15]. Signal modelling starts from the spectrum. Only local maxima or spectral peaks are considered to be representative of the frequency components present in the signal.

As a result, the pattern is then composed of the sum of a limited number of sinusoids k.

$$x(t) = \sum_k \alpha_k e^{2\pi i f_0 kt + i \phi_k}. \tag{1}$$

The model parameters are, respectively, the amplitudes α_k, frequencies $2\pi k f_0$ and the phase ϕ_k [13].

This basic form of the sinusoidal model is for strictly stationary signals, composed of a sum of sinusoids whose parameters are fixed over the considered time portion [3].

Such representation is based of course on the nature of the vibration signal. In our case in speech production, a vibration of the vocal cords is the source of voiced sounds in which this model is locally well suited [2, 10].

To compute the parameters exposed in (1), we use the short time Fourier transform.

$$F_x(t, f) = \int x(\tau)h(\tau - t)e^{-2\pi i f \tau}d\tau. \tag{2}$$

For the purpose of illustration, we will consider a signal with the following characteristics: a vowel "a" pronounced by a woman with a sampling frequency of 16,000 Hz.

The spectre is computed through the STFT in (2) using a hamming window with the same length as the signal.

The spectrogram, squared magnitude of the STFT $S_x(t, f) = |F_x(t, f)|^2$, is represented using a hamming window with same length as the signal and 50 % of overlapping (Figs. 1 and 2).

We then extract the local maxima from the spectrum (α_i). We have considered until the 14th coefficient because the information in the spectrum after that rank has

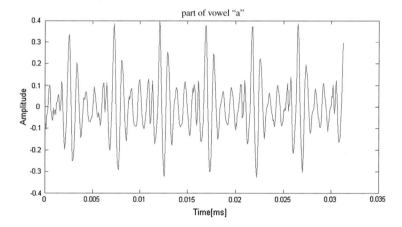

Fig. 1 Signal: part of vowel "a"

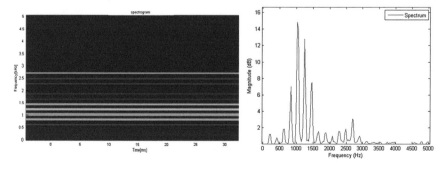

Fig. 2 Spectrum and spectrogram of the signal

Table 1 α_i values

alpha 1	alpha 2	alpha 3	alpha 4	alpha 5	alpha 6	alpha 7
1,4070	0,8153	1,8820	6,9970	14,6800	12,6300	7,3770

Table 2 α_i values

alpha 8	alpha 9	alpha 10	alpha 11	alpha 12	alpha 13	alpha 14
1,5540	1,4200	0,9844	1,5580	1,8530	3,0120	0,9532

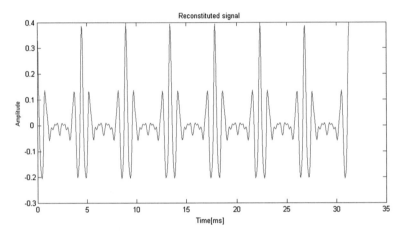

Fig. 3 Reconstituted signal

no significant impact. We also compute the fundamental frequency ($f_0 = 224$ Hz) (Tables 1 and 2).

To reconstruct the voiced speech signal we need to add the α parameters above and f_0 in (1). The resulting signal have the same period as the real signal but with an amplitude significantly high compared to it. To circumvent this difference we need to add a factor to adjust the signal amplitude. This factor is computed using the ratio between the real signal and the model. At the end we can generate this signal (Fig. 3).

3 Wigner Ville Transform

3.1 Definition and Major Properties

The WVT $W_x(t, f)$ of a time signal $x(t)$ is expressed by [14]:

$$W_x(t, f) = \int_{-\infty}^{\infty} x(t + \frac{\tau}{2}) x^*(t - \frac{\tau}{2}) e^{-2\pi i f \tau} d\tau. \tag{3}$$

The WVT can be considered as a suitable choice to proceed with the time and frequency distribution as it is correlating the signal with a time- and frequency translated version of itself without using a windowing function like the ones used in the Fourier and wavelet transforms [8, 12].

This particular property exempt WVT from the blurry effect due to the windowing function, and as a result, the WVT gives the representation that has the best accuracy in the time frequency [19].

At this point, the WVT may seem to be a powerful tool for signal analysis [11]. However, it misses a fundamental property as an energy density which is positivity and the interferences created by the quadratic aspect of this transform [12].

3.2   WVT of the Speech Signal Model

First, we need to express the WVT with a voiced speech signal model to get the self and cross-terms expressions.

WVT of a speech signal can be written, based on (3) and the signal $x(t) = \sum_k \alpha_k e^{2\pi i f_0 k t}$. as:

$$W_x(t, f) = \sum_q \sum_p \alpha_p \alpha_q e^{2\pi i (p-q) f_0 t} \delta(f - f_0 \frac{(p+q)}{2}). \tag{4}$$

Equation 4 can be written as:

$$W(t, f) = \sum_k W_{kk}(t, f) + \sum_{k \neq p} W_{kp}(t, f).$$

When $p = q$

$$W_x(t, f) = \sum_k \alpha_k^2 \delta(f - k f_0). \tag{5}$$

when $p > q$ as $W_{ij} * (t, f) = W_{ji}(t, f)$ we write

$$W(t, f) = \sum_k W_{kk}(t, f) + \sum 2 \Re W_{kp}(t, f).$$

$$W_x(t, f) = \sum_{p>q} 2 \alpha_p \alpha_q \cos(2\pi (p-q) f_0 t) \delta(f - f_0 \frac{(p+q)}{2}). \tag{6}$$

where $W_{kk}(t, f)$ are self-terms energy distribution of $x_k(t)$ and $W_{kp}(t, f)$ are cross-terms of WVT. The presence of non-negligible cross-terms, resulting from interactions between signal components, may lead to an inaccurate visual interpretation of the signals time-frequency distribution, since they may overlap with the self-terms of energy distribution.

Computing with phase will not change the result:

$$x(t) = \sum_k \alpha_k e^{2\pi i f_0 kt + i\phi_k}.$$

$$W_x(t, f) = \sum_q \sum_p \alpha_p \alpha_q e^{2\pi i (p-q) f_0 t + i(\phi_p - \phi_q)} \delta(f - f_0 \frac{(p+q)}{2}).$$

When $p = q$

$$W_x(t, f) = \sum_k \alpha_k^2 \delta(f - kf_0)$$

when $p > q$

$$W_x(t, f) = \sum_{p>q} 2\alpha_p \alpha_q \cos(2\pi(p-q) f_0 t + (\phi_p - \phi_q)) \delta(f - f_0 \frac{(p+q)}{2}).$$

We will then use the model we developed and the parameters extracted from the spectrum to represent the WVT of the voiced speech model. First, we will inject these values in (5) and (6). The WVT representation is (Fig. 4).

The Wigner Ville Transform of the real signal (Fig. 5).

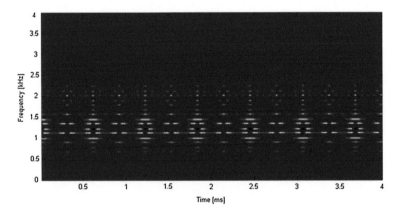

Fig. 4 Wigner Ville Transform of the voiced speech model

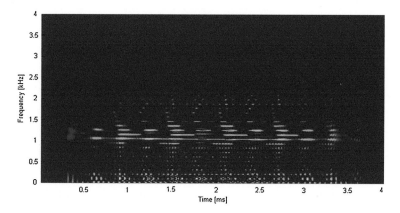

Fig. 5 Wigner Ville Transform of the real signal

3.3 Analysis

The second step is superposing the two representations to identify similarities and differences. Both representation of the WVT have the following characteristics: through (5) and (6) we can notice that the self-terms are located in kf_0 and the cross-term are located in $f_0\frac{(p+q)}{2}$. For $\frac{(p+q)}{2} = k$ the cross-terms overlap the self-terms, which leads to a problem of energy analysis.

For the same frequency, amplitude variation is due to the product $\alpha_p * \alpha_q$, and the presence of periodicity variation and negative values are due to the cosine in (6).

The two representations don't look exactly the same and the reason is the real aspect of the signal which is non stationary and the energy variation between the real and the reconstituted signal.

Let's point out that the energy distribution is accurate in the WTV whereas it is blurred in the spectrogram.

Apart the accuracy of energy, the WVT is characterized by the interferences (cross terms) Eq. 6. Several studies worked on reducing the resulting interferences after the WVT by the use of a smoothing function [4, 5, 9, 17].

The WVT of the window is a smoothing function that leads to the spectrogram. The spectrogram of the signal $x(t)$ can be estimated by computing the squared magnitude of $F(t, f) : S(t, f) = |F(t, f)|^2$ but also [6]:

$$S(t, f) = \int \int W_x(\tau, v) W_t(\tau - t, v - f) dt df.$$

The spectrogram of the signal can be written as:

$$F_x(t, f) = X(f) * e^{-2\pi i f t} W_t(f) \quad with \quad X(f) = \sum_{k=0}^{N} \alpha_k \delta(f - kf_0)$$

$$F_x(t, f) = X(f) * e^{-2\pi i f t} W_t(f) = \sum_{k=0}^{N} \alpha_k \delta(f - kf_0) * e^{-2\pi i f t} W_t(f)$$

$$F_x(t, f) = \sum_{k=0}^{N} \alpha_k e^{-2\pi i (f - kf_0)t} W_t(f - kf_0)$$

$$S_x(f, t) = \sum_{p=0}^{N} \sum_{q=0}^{N} \alpha_p \alpha_q e^{-2\pi i t (f - (p-q)f_0)} W_t(f - pf_0) W_t(f - qf_0)$$

After reducing the interferences (smoothing) these methods may distort the energy accuracy. By studying the WVT of voiced speech signal model more closely, we can figure out how to improve the Wigner Ville representation.

3.4 Wigner Ville of Periodic Signal

We can extend our previous work for periodic signal by using a window.
The WVT of the voiced speech signal model can be written as:

$$W_x(t, f) = \sum_{q} \sum_{p} \alpha_p \alpha_q e^{2\pi i (p-q) f_0 t} \int_{-\infty}^{\infty} e^{-2\pi i (f - (p+q)\frac{f_0}{2})\tau} d\tau$$

this expression is equivalent to the Eq. 4.
In the case of a rectagular window

$$X(f) = \int_{-T}^{T} e^{-2\pi i (f - (p+q)\frac{f_0}{2})\tau} d\tau = 2T \sin c(2\pi f T - (p+q)\pi)$$

$$W_x(t, f) = \sum_{q} \sum_{p} \alpha_p \alpha_q e^{2\pi i (p-q) f_0 t} 2T \frac{\sin(2\pi(f - (p+q)\frac{f_0}{2})T)}{2\pi(f - (p+q)\frac{f_0}{2})T}$$

when $p = q$

$$W_x(t, f) = \sum_{k} \alpha_k^2 2T \frac{\sin(2\pi(f - kf_0)T)}{2\pi(f - kf_0)T}$$

when $p > q$

$$W_x(t, f) = \sum_{p>q} 2\alpha_p \alpha_q \cos(2\pi i (p-q) f_0 t) 2T \frac{\sin(2\pi(f - (p+q)\frac{f_0}{2})T)}{2\pi(f - (p+q)\frac{f_0}{2})T}$$

4 Conclusion

In this paper, we used the Wigner Ville Transform in two ways: (i) represent the energy distribution of a voiced speech signal model (this expression was made by writing the Wigner Ville formula with model speech model, and injecting the signals parameters in this formula); and (ii) represent the WVT of a real voiced speech signal. The purpose of this study is to examine the Wigner Ville transform from a theoretical point of view.

Through creating this mathematical expression of the WVT of a voiced speech signal model we were able to emphasize on some of the aspects of this transform: energy accuracy, amplitude, periodicity, self-terms and cross-terms.

We concluded that the WVT is a very interesting representation for the voiced speech signal, except for the interferences.

In future works we will work on reducing the interferences to have the best energy representation of speech signal. We will focus, also on the WVT of non-stationary signals as a generalization of this work [7, 18].

References

1. Boashash, B.: Time-Frequency Signal Analysis and Processing. A Comprehensive Reference. Elsevier (2003)
2. Calliope: La parole et son traitement automatique. Collection technique et scientifique des telecommunications (1989)
3. Ellis, D.P.W.: An introduction to signal processing for speech. In: The Handbook of Phonetic Sciences (2008)
4. Auger, F., Chassande-Mottin, E., Patrick, F.: On the statistics of spectrogram reassignment vectors. Multimedia Syst. Signal Process. 9 (1998)
5. Meyer, G., Plante, F., Ainsworth, W.A.: Improvement of speech spectrogram accuracy by the method of reassignment. IEEE Trans. Speech Audio Process. 6(3) (1998)
6. Meyer, G., Plante, F., Ainsworth, W.A.: Speech signal analysis with reallocated spectrogram. IEEE (1994)
7. Flandrin, P.: Time-frequency representation of non-stationary signals. Traitement du signal 6(2) (1989)
8. Flandrin, P.: Remarks on the notion of time-frequency localization. Traitement du signal 15(6) (1998)
9. Flandrin, P., Auger, F.: Improving the readability of time-frequency and time-scale representations by the reassignment method. IEEE Trans. Signal Process. 43(5), (1995)
10. OShaughnessy, D., Deng, L.: A dynamic and optimization-oriented approach. Speech Process. Marcel Dekker Inc. New York, NY, U.S.A. (2003)
11. Stankovic, S., Stankovic, L., Dakovic, M.: From the stft to wigner distribution. IEEE Signal Process. Mag. 31(3), 163–174 (2014)
12. Mallat, S.: A Wavelet Tour of Signal Processing, 2nd edn. Academic Press (1998–1999)
13. Gerkmann, T., Krawczyk, M.: STFT phase reconstruction in voiced speech for an improved single-channel speech enhancement. IEEE/ACM Trans. Audio Speech Lang. Process. 22(12), 1931–1940 (2014)
14. Escudi, B., Flandrin, P.: Principle and application of time-frequency analysis by means of the wigner-ville transform. Traitement du signal 2(2) (1985)

15. Quatieri, T.F., McAulay, R.J.: Speech processing based on a sinusoidal model. Lincoln Lab. J. **1**(2) (1988)
16. Portnoff, M.R.: Implementation of the digital phase vocoder using the fast Fourier transform. IEEE Trans. Acoust. Speech Signal Process. ASSP **2**(3) (1976)
17. Macleod, M.D., Hainsworth, S.W.: Time frequency reassignment: a review and analysis. Technical Report, Cambridge University Engineering Department, CUED/F-INFENG/TR.459
18. Flandrin, P., Martin, W.: Wigner Ville spectral analysis of non stationary process. IEEE Trans. Acoust. Speech Audio Process. ASSP **33**(6), (1985)
19. Goutte, R., Zhu, Y.M., Peyrin, F.: Transformation de Wigner-Ville: description d'un nouvel outil de traitement du signal et des images. Ann. Telecommun. (1987)

Glottal Closure Instant Detection by the Multi-scale Product of the Derivative Glottal Waveform Signal

Ghaya Smidi, Aicha Bouzid and Noureddine Ellouze

Abstract This paper is about the detection of the glottal closure instants (GCI) by the multi-scale product (MP) of the derivative glottal waveform signal. Based on the source filter model, the derivative glottal waveform signal is estimated by the inverse filtering of the non pre-emphasized speech signal with the LP coefficients. The derivative glottal waveform signal represents the real excitation of the vocal tract and shows discontinuities at GCI. MP acts as a discontinuity detector. A preprocessing step is added to improve the GCI detection. The performance of our method is evaluated on the Keele university database and compared to the MP applied directly on the speech signal. Using the preprocessing phase, the MP applied on the derivative glottal waveform signal gives an identification rate of 99.21 % and an accuracy to ± 0.25 ms of 87.32 % versus an identification rate of 99.15 % and an accuracy to ± 0.25 ms of 75.78 % for the MP method applied directly on speech signal.

Keywords Glottal closure instant (GCI) · Multi-scale product (MP) · Derivative glottal waveform signal

1 Introduction

Quasi-periodic vibrations of the vocal cords form the essential excitation of the vocal tract in a voiced speech. These vibrations are the result of a glottis behavioral phenomenon, caused by the air pressure coming from the lungs. In the larynx cycle, the instants of the opening and closing of the glottis are important. These moments are noted GCI for closing (Glottal Closure Instant) and GOI for opening (Glottal Opening

G. Smidi (✉) · A. Bouzid · N. Ellouze
Signal, Image and Information Technology Laboratory, ENIT, Tunis, Tunisia
e-mail: Smidi_ghaya@yahoo.fr

A. Bouzid
e-mail: bouzidacha@yahoo.fr

N. Ellouze
e-mail: nourellouze@yahoo.fr

© Springer International Publishing Switzerland 2016   191
A. Esposito et al. (eds.), *Recent Advances in Nonlinear Speech Processing*,
Smart Innovation, Systems and Technologies 48,
DOI 10.1007/978-3-319-28109-4_19

Instant). The major excitation of the vocal tract occurs at GCI. The vocal cords are abruptly closed and stay closed for approximately the half of the cycle (closed phase), the airflow stops quickly, resulting in a discontinuity of the waveform of the glottal flow. Thus the GCI is the most significant moment when a periodic reference point is required in the larynx cycle.

The estimation of the pitch contour and the boundaries of each larynx cycles have been recently allowed by the detection of GCI. This information was used in the anticausal-causal deconvolution of the speech signal [1], prosodic speech modification [2], dereverberation of the speech [3], glottal flow estimation [4] and so on. The critical importance of GCI in various fields, has led researchers to work on GCI detection methods directly from the speech signal.

Many methods have been developed to detect these crucial instants from the speech waveform; besides a quantitative review has recently been developed by Drugman and et al. to highlight five of the most effective methods determining the GCIs from the speech signal namely the Hilbert Envelope-based method (HE) [5], the Zero Frequency Resonator-based method (ZFR) [6], DYPSA [7], SEDREAMS [8] and the YAGA algorithm [9]. In the YAGA algorithm, GCI are detected by the application of the multi-scale product (MP) method on the derivative glottal waveform signal then the application of the group delay method followed by the dynamic programming.

In this paper, we focus on the application of the MP method on the derivative glottal wave form signal without using the group delay method and dynamic programming. The suggested method, the MP applied on the derivative glottal wave form signal, is compared to the MP method applied directly on the speech signal.

This paper is structured as follows. Section 2 deals with the discontinuity detection of the voiced speech by the MP method. Section 3 presents GCI detection by the MP of the derivative glottal waveform signal. In Sect. 4, results obtained on the Keele database are shown and compared with MP applied directly on the speech signal. A conclusion is given in Sect. 6.

2   Discontinuity Detection of Voiced Speech by the Multi-scale Product Method

The multi-scale product (MP) method is the product of the coefficients of the wavelet transform (WT) at some scales. The MP of a signal f(n) is defined by Eq. (1):

$$p(n) = \prod_j w_{s_j}[\mathrm{f}(n)] \tag{1}$$

where $w_{s_j}[\mathrm{f}(n)]$ is the WT of the signal f(n) at the scale s_j.

The WT is considered as a multi-scale differential operator with n order of the smoothed signal.

It has been shown that for relatively fine scales, the WT of the signal presents maxima at its rapid transitions. However, the maxima obtained by the WT are degraded by noise at low levels and generally smoothed with strong scales [10]. The MP solves this problem and hence is considered as one of the most recent methods that has proved its efficiency and robustness for GCI detection [11].

3   The Proposed Method

Our method is essentially based on the estimation of the derivative glottal waveform signal and the application of the MP on this obtained signal

3.1   Estimation of the Derivative Glottal Waveform Signal Based on the Source Filter Model of Speech Production

Relying on the source filter model, a voiced speech sound can be modeled as the convolution of the derivative glottal waveform signal noted $u'(n)$, and the transfer function of the vocal tract noted $v(n)$.

$$s(n) = u'(n) * v(n) \tag{2}$$

In the z domain, the Eq. (2) takes the following form:

$$S(z) = (1 - z^{-1})U(z) \cdot V(z) \tag{3}$$

Hence the derivative glottal waveform signal is given by:

$$(1 - z^{-1})\widehat{U}(z) = S(z)/\widehat{V}(z) \tag{4}$$

According to the Eq. (4), the estimated derivative glottal waveform signal, is simply the inverse filtering of the speech signal by the estimated transfer of the vocal tract $\widehat{V}(z)$.

In the literature, $v(n)$ is considered as an all pole transfer filter [12]. In z domain we have:

$$V(z) = \left(1 + \sum_{k=1}^{p} a_k z^{-k}\right)^{-1} \tag{5}$$

Replacing $V(z)$ by its expression in (5), we obtain:

$$S(z)\left(1 + \sum_{k=1}^{p} a_k z^{-k}\right) = (1 - z^{-1}) \cdot U(z) \tag{6}$$

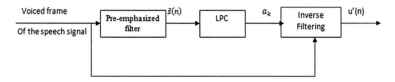

Fig. 1 Estimation of the derivative glottal waveform signal u′(n)

In the time domain, the expression becomes:

$$s(n) = -\sum_{k=1}^{p} a_k \, s(n-k) + u'(n) \qquad (7)$$

The first term is a p-order LPC of the signal s(n), where a_k are the coefficients of the transfer function of the vocal tract. The second term is related to the derivative glottal waveform signal as the residue of the LPC decomposition of the speech signal s(n).

The V(z) is determined through the estimation of a_k coefficients by the autocorrelation method. This method requires the use of a pre-emphasized high pass filter [9].

Figure 1 presents the estimation of the derivative glottal waveform signal u′(n).

3.2 Detection of GCI by the Multiscale Product of the Derivative Glottal Waveform Signal

The speech signal characterized by a non-stationary nature, estimating LPC coefficients is performed in a sliding window. Each voiced speech segment is decomposed into frames, on which we apply the pre-emphasis filter and a Hanning window. The prediction coefficients are calculated and thus the transfer of the vocal tract is determined. Inverse filtering of the non-pre-emphasized speech signal, allows then to determine the residual signal in the analysis window. The derivative glottal waveform signal is finally generated by the overlap add method for the entire voiced segment. The derivative glottal waveform signal represents the real excitation of the vocal tract and shows discontinuities at GCI.

The WT is then applied as this signal at three scales. The product of these three WT results in the MP of the derivative glottal waveform signal (Fig. 2).

3.3 Improving the GCI detection method

The windowing at the beginning and the end of the segment makes locating GCI relatively difficult because of the low amplitude maxima. Therefore a processing

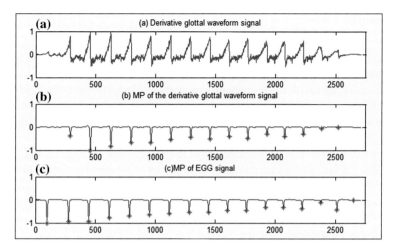

Fig. 2 Example of GCI extraction on the 55th voiced segment for speaker m2: **a** the derivative glottal waveform signal, **b** MP of the derivative glottal waveform signal, **c** the MP of EGG signal

of the voiced signal segment, consisting in adding L/2 zeros at the beginning and end of the segment is proposed, where L is the length of the analysis window. The calculation of the derivative glottal waveform signal is conducted in each window then the overlap add method is used to calculate this signal throughout the entire voiced segment. At the end of the processing, the base signal is extracted (Fig. 3).

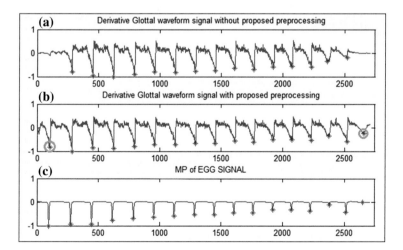

Fig. 3 a Derivative glottal waveform signal without the proposed preprocessing. **b** Derivative glottal waveform signal with the proposed preprocessing. **c** The MP of EGG signal

4 Results

The experiments presented in this section have been carried out on the Keele university database consisting in 10 speakers: 5 men and 5 women, labeled m1, m2, m3, m4, m5 and f1, f2, f3, f4, f5: each speaker pronounces the text "the north wind story" for about 40 s per speaker in an anechoic chamber. The EGG signal is recorded simultaneously with the speech signal in the same conditions. Both signals are sampled at 20 kHz and coded by 16 bits.

The proposed method is compared with the MP applied directly on the speech signal. The GCI references are extracted by applying MP method on the EGG signal.

The EGG signal needs to be time-aligned with the audio signal. In this work, the time alignment is done by calculating the maximum of the cross-correlation function.

A 24th order LP analysis is performed on a Hanning window of 512 samples shifted every 64 samples. The first derivative of a Gaussian function is used as the WT and the following scales are used to generate the MP of $u'(n)$ signal. $s1 = 1$, $s2 = 1.2$ and $s3 = 1.5$.

The proposed method is evaluated according to the following measures:

- Identification Rate (IDR): the proportion of larynx cycles for which a unique GCI is detected.
- Missing Rate (MR): the proportion of larynx cycles for which no GCI is detected.
- False Alarm Rate (FAR): the proportion of larynx cycles for which more than one GCI is detected.
- Identification Accuracy (IDA): the standard deviation of the timing error distribution.
- The accuracy to ±0.25 ms: the rate of detections for which the timing error is smaller than this bound.

Table 1 shows the evaluation and the comparative results of GCI detection on the Keele database in terms of identification rate, missing rate, false alarms rate

Table 1 Comparative results in terms of Identification Rate (IDR), Miss Rate (MR) and False Alarm Rate (FAR)

Speaker	Method	IDR (%)	MR (%)	FAR (%)	IDA (ms)	Accuracy to ±0.25 ms
Keele database	MP(s(n))	99.15	0.60	0.26	0.33	75.78
	MP(u'(n)) without pre-processing	98.76	1.04	0.21	0.28	87.76
	MP(u'(n)) with pre-processing	99.21	0.58	0.22	0.28	87.32

Identification Accuracy (IDA) and accuracy to ±0.25 ms

identification accuracy (IDA) and accuracy to ± 0.25 ms, characterizing the error probability densities.

Table 1 shows that the MP applied on the signal u$'$(n) calculated with the proposed preprocessing step presents the best reliability and practically the same accuracy as the MP applied on u$'$(n) signal calculated without preprocessing. The preprocessing phase reduces visibly the rate of missed GCIs.

5   Computational Complexity of GCI Detection Methods

To compare the computational complexity of GCI detection methods evaluated in this paper, we propose to calculate the Relative Computation Time (RCT) for each one of them:

$$RCT(\%) = 100 \cdot \frac{CPU\ time\ (s)}{Sound\ duration(s)}$$

The averaged Relative Computation Time (RCT), obtained for our Matlab implementations for all keele database speakres, with an Intel (R) Core(TM) i3-3217 u cpu @1.80 GHZ with 4GO of RAM.

Table 2 shows Relative Computation Time (RCT), in % for evaluated methods averaged across all speakers Keele database.

We note that the MP applied on speech signal is the fastest method (with RTC around 3.70 %) followed by MP applied on derivative glottal waveform signal without preprocessing phase (with RTC around 24.64 %) and finally MP applied on derivative glottal waveform signal with preprocessing phase (a RTC around 50.14 %). This latter is the heaviest because of the execution time given to the pre-treatment phase.

Table 2 Relative Computation Time (RCT) for evaluated methods averaged across all speakers Keele database

Method	CPU (s)	RCT (%)	Sound duration of all Keele database (s)
MP(s(n))	12.48	3.70	337.12
MP(u$'$(n)) without preprocessing	83.08	24.64	
MP(u$'$(n)) with preprocessing	169.04	50.14	

6 Conclusions

This paper focuses on the detection of the glottal closure instants based on the Multiscale product of the derivative glottal waveform signal. The measures given by the proposed method are valued on the Keele University database. An improving of the detection method is brought through a preprocessing step. The results shows that the MP of the derivative glottal waveform signal calculated with preprocessing phase presents the best performance; an identification rate which exceeds 99 % and an accuracy to ±0.25 ms which exceeds 87 %.

References

1. Bozkurt, B., Dutoit, T.: Mixed-phase speech modeling and formant estimation, using differential phase spectrums. In: ISCA ITRW VOQUAL03, pp. 21–24 (2003)
2. Moulines, E., Charpentier, F.: Pitch-synchronous waveform processing techniques for text-to-speech synthesis using diphones. Speech Commun. **9**(5–6), 453–467 (1990)
3. Gaubitch, N.D., Naylor, P.A.: Spatiotemporal averaging method for enhancement of reverberant speech. In: Proceedings of IEEE International Conference on Digital Signal Processing (DSP), Cardiff, UK (2007)
4. Wong, D.Y., Markel, J.D., Gray, J.A.H.: Least squares glottal inverse filtering from the acoustic speech waveform. IEEE Trans. Acoust. Speech Signal Process. **27**(4), 350–355 (1979)
5. Rao, K.S., Prasanna, S.R.M., Yegnanarayana, B.: Determination of instants of significant excitation in speech using Hilbert envelope and group delay function. IEEE Signal Process. Lett. **14**(10), 762–765 (2007)
6. Murty, K.S.R., Yegnanarayana, B.: Epoch extraction from speech signals. IEEE Trans. Audio Speech Lang. Process. **16**(8), 1602–1613 (2008)
7. Naylor, P., Kounoudes, A., Gudnason, J., Brookes, M.: Estimation of glottal closure instants in voiced speech using the DYPSA algorithm. IEEE Trans. Audio Speech Lang. Process. **15**(1), 34–43 (2007)
8. Drugman, T., Dutoit, T.: Glottal closure and opening instant detection from speech signals. In: Proceedings of Interspeech Conference (2009)
9. Thomas, M.R.P., Gudnason, J., Naylor, P.A.: Estimation of glottal closing and opening instants in voiced speech using the YAGA algorithm, Feb (2012)
10. Bouzid, A., Ellouze, N.: Produit multiéchelle pour la détection des instants d'ouverture et de fermeture de la glotte sur le signal de parole. JEP (2006)
11. Tuan, V.N., d'Allessandro, C.: Robust glottal closure detection using the wavelet transform. In: Proceedings of the European Conference on Speech Technology, pp. 2805–2808 (1999)
12. Markel, J.D., Gray Jr., A.H.: Linear Prediction of Speech. Springer (1976)

Non-linear Dynamics Characterization from Wavelet Packet Transform for Automatic Recognition of Emotional Speech

J.C. Vásquez-Correa, J.R. Orozco-Arroyave, J.D. Arias-Londoño,
J.F. Vargas-Bonilla and Elmar Nöth

Abstract A new set of features based on non-linear dynamics measures obtained from the wavelet packet transform for the automatic recognition of "fear-type" emotions in speech is proposed. The experiments are carried out using three different databases with a Gaussian Mixture Model for classification. The results indicate that the proposed approach is promising for modeling "fear-type" emotions in speech.

Keywords Non-linear dynamics · Non-linear speech processing · Speech emotion recognition · Wavelet packet transform

1 Introduction

Speech is the main process of communication between humans. This fact has motivated researches to use it as a mechanism of interaction between humans and computers. The challenge now is not only to recognize the words and sentences but also the paralinguistic aspects of speech such as emotions and personality of the speaker. In the last few years the interest of the research community has been focused on the detection of "fear-type" emotions such as anger, disgust, fear, and desperation, which appear in abnormal situations when the human integrity is at risk [1]. One of the main aims of speech analysis is to find suitable speech features to represent the emotional state of a speaker. In related works, the characterization has been focused on prosodic features, spectral and cepstral features such as Mel Frequency Cepstral Coefficients (MFCC), and voice quality features such as noise measures [1]. In [2] the authors use Berlin [3], and enterface05 [4] databases for emotion recognition. They use MFCC joint to their first and second derivatives, and perform the classification using a Deep Neural Network with a Hidden Markov Model (DNN-HMM). The

J.C. Vásquez-Correa · J.R. Orozco-Arroyave (✉) · J.D. Arias-Londoño · J.F. Vargas-Bonilla
Faculty of Engineering, Universidad de Antioquia UdeA, Medellin, Colombia
email: rafael.orozco@i5.informatik.uni-erlangen.de

J.R. Orozco-Arroyave · E. Nöth
Pattern Recognition Lab, Friedrich-Alexander-Unversität, Erlangen, Germany

© Springer International Publishing Switzerland 2016   199
A. Esposito et al. (eds.), *Recent Advances in Nonlinear Speech Processing*,
Smart Innovation, Systems and Technologies 48,
DOI 10.1007/978-3-319-28109-4_20

reported accuracies are 77.92, and 53.89 % for Berlin, and enterface05 databases, respectively. In [5] the authors use enterface05 [4], and FAU Aibo [6] databases for emotion recognition. They use acoustic features related to MFCC, energy, and fundamental frequency, and propose a method based on least square regression (LSR) for recognition. The reported accuracies are 69.33 % in enterface05, and 60.50 % in Aibo database, respectively.

On the other hand, the use of Non-Linear Dynamics (NLD) measures in speech processing tasks has been increased in the last years. In [7] the authors perform experiments in three different databases, including the Emotional prosody speech and transcripts of the Linguistic Data Consortium (LDC) [8], the Berlin database [3], and the Polish emotional speech database [9]. The authors characterize the speech recordings using features related to NLD and perform the classification of different emotional states using an artificial neural network. The reported accuracies are 80.75, 75.40, and 72.78 % for each database, respectively. In [10] the authors use Berlin, and SUSAS [11] databases to evaluate the representation capability of the Hurst exponent (HE) obtained from the Discrete Wavelet Transform (DWT) to recognize different emotions and to detect stress from speech. The author perform segmentation between voiced and unvoiced segments, and calculate the features only for voiced segments. The speech signals are modeled using a Gaussian Mixture Model (GMM), and the reported accuracies are 68.1 and 64 %, for the Berlin and SUSAS databases, respectively. In [12] the authors use SUSAS [11] database for automatic stress recognition in speech. They use features related to energy and entropy obtained from Wavelet Packet Transform (WPT). Automatic recognition is performed by means of a Linear Discriminant Analysis (LDA), and the reported accuracy is about 91 %.

In previous works, we calculate acoustic features obtained from WPT [13]. In this paper, we propose a new set of features related to NLD obtained from WPT for fear-type emotion recognition in speech signals. WPT provides a time-frequency representation in different resolutions and the NLD features are calculated on each decomposed band. The features are calculated on speech recordings of three different databases very used in emotion recognition: (i) Berlin [3], (ii) GVEESS [14], and (iii) enterface05 [4]. Classification is performed using a GMM derived from a Universal Background Model (GMM-UBM). The rest of paper is distributed as follows: Sect. 2 contains the description about the characterization and classification processes. Section 3 describes the experimental framework, the databases, and the obtained results. Finally, Sect. 4 includes the conclusions derived from this study.

2   Materials and Methods

Figure 1 shows the general scheme of the proposed methodology. It consists of four stages. First the voiced and unvoiced segments of speech are separated using the software Praat [15] in order to analyze features estimated from each one of them. Second the wavelet decomposition is performed on each segment separately. Third each decomposed band is characterized separately as follows: for the voiced segments

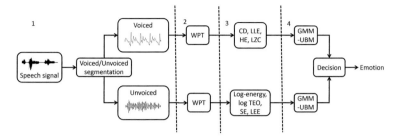

Fig. 1 General scheme of the methodology

four NLD features are calculated including Correlation Dimension (CD), Largest Lyapunov exponent (LLE), HE, and Lempel-Ziv complexity (LZC); and for the case of unvoiced segments another four features are estimated including the log energy, the log energy derived from Teager Energy Operator (TEO), the Shannon Entropy (SE), and the Log Energy Entropy (LEE). The difference in the features estimated both for voiced and unvoiced segments is based on the fact that features estimated for voiced segments are related to perturbation of the fundamental frequency, and the excitation source [10]. These features cannot be estimated for unvoiced segments. Finally, the fourth stage of the methodology includes the GMM-UBM modeling. The decision of which emotion is present on each recording is taken by the combination of the posterior probabilities produced by the classifiers applied on the voiced and unvoiced feature vectors.

2.1 Feature Estimation

Taken's embedding and phase space: the NLD analysis begins with the reconstruction of the phase space of the speech signal according to the embedding process [7]. A time series $x(i)$ $i = 1, 2, \cdots N_m$, can be represented in a new space which is defined as $X[k] = \{\mathbf{x}[k], \mathbf{x}[k + \tau], \mathbf{x}[k + 2\tau], \mathbf{x}[k + (m - 1)\tau]\}$. $N_m = N - (m - 1)\tau$ is the reconstructed vector length, τ is the time delay, and m is the embedding dimension.

Correlation Dimension (CD): this feature allows the estimation of the exact space that is occupied by the reconstructed vector in the phase space. It is an indicator about the complexity and dimensionality of speech signal [7].

Largest Lyapunov Exponent (LLE): this measure quantifies the exponential divergence of neighbor trajectories in a phase space. This measure provides an indicator of the aperiodicity of a speech signal [7].

Hurst Exponent (HE): this feature expresses the long term dependence of a time series. HE is defined according to the asymptotic behaviour of the rescaled range of a time series as a function of a time interval. HE can be used to represent the emotional state of speech according to the arousal level of the signal [10].

Lempel Ziv Complexity (*LZC*): this feature establishes a measure about the degree of disorder of spatio-temporal patterns in a time series. The LZC reflects the rate of new patterns in the sequence; and ranges from 0 (deterministic sequence) to 1 (random sequence). LZC distribution shows values nearer to 1 for fear and anger speech, than in case of neutral speech [7].

log-Energy: it is calculated according to the Eq. 1 [16].

$$E(k) = log \left[\frac{\sum_{l=1}^{N_k} x(l)^2}{N_k} \right] \qquad (1)$$

log-Energy of TEO: the TEO was developed in order to measure the changes in speech signal produced under stress. The TEO of a signal is calculated as $TEO(x) = x(k)x(k)^* - x(k+1)x(k-1)$. Finally, the log Energy of TEO is calculated according to the Eq. 2 [16].

$$E_{TEO}(k) = log \left[\frac{\sum_{l=1}^{N_k} |x(l)x(l)^* - x(l+1)x(l-1)|}{N_k} \right] \qquad (2)$$

Shannon and log energy entropy: it models the complexity of a system. In this paper we estimate two entropy measurement, the Shannon entropy, and the log energy entropy. These features are calculated according to [12] as $H1(k) = -\sum_{l=1}^{n} x(l)^2 log(x(l)^2)$ and $H2(k) = -\sum_{l=1}^{n} log(x(l)^2)$, respectively. n is the number of bins used to estimate the probability density function of the wavelet decomposition.

2.2 Classification

The features extracted from voiced and unvoiced segments were classified separately using a GMM adapted from a UBM, using Maximum A Posterior (MAP) to derive a specific GMM from the UBM [17]. A GMM can be defined as a probabilistic model represented by the sum of several multivariate Gaussian components. The model is expressed according to its probability density function.

$$p(\mathbf{x}|\Theta) = \sum_{k=1}^{M} P_k \mathcal{N}(\mathbf{x}|\mu_k, \Sigma_k) \qquad (3)$$

where M is the number of Gaussian components, P_k is the prior probability (mixing weight), and $\mathcal{N}$ is a multivariate Gaussian density function. The UBM is trained using the Expectation Maximization (EM) algorithm [17] using a population of all classes. Then the specific GMM for each class is adapted using the MAP method.

Finally, given a sample $X = \mathbf{x}_1, \mathbf{x}_2, \ldots, \mathbf{x}_T$, where $\mathbf{x}_i$ is the feature vector extracted from the frame i, the decision about to which class belongs each speech sample is taken evaluating the maximum Log-Likelihood according to the Eq. 4.

$$LL(X, \Theta) = \sum_{t=1}^{T} log(p(\mathbf{x}_t|\Theta)) \tag{4}$$

The posterior probabilities obtained from the GMM model based on voiced segments is combined with those obtained from the GMM model based on unvoiced segments. Both probabilities are normalized according to the length of the frames to avoid possible bias introduced due to the length-dependence of the models. The combined probability is defined as $LL_{comb} = \alpha LL(X, \Theta_{Voiced}) + (1 - \alpha)LL$ $(X, \Theta_{Unvoiced})$, $\alpha \in (0.1, 0.2, \ldots, 0.9)$ according to [18], and it is a weight coefficient which is optimized on test.

3 Experimental Framework and Results

3.1 Experimental Setup

All experiments were performed using leave one group speaker out cross-validation (LOGSO-CV), with different numbers of Gaussian components for classifier (from 2 to 8) with diagonal covariance matrix. The value of α was optimized through a grid search with $0 < \alpha < 1$. The selection criteria was based on the obtained accuracy on the test set. The length of frames is selected according to the sample frequency in order to guarantee enough number of points for the wavelet decomposition. Frames of 1764 samples are selected, which is equivalent to 40 ms in cases when $fs = 44100$ Hz, and 110 ms when $fs = 16000$ Hz [13]. In both cases an overlapping percentage of 50 % is used. All of the coefficients from the second and third level of WPT are considered. Daubechies3 is used as the mother wavelet. The experiments were carried out to recognize different "fear-type" emotions in speech, such as anger, disgust, fear and desperation.

3.2 Databases

- Berlin emotional database [3]: it contains 534 voice recordings produced by 10 speakers who acted 7 different emotions. The recordings were sampled at 16 kHz. In this paper three of the seven emotions of the database are considered for the automatic recognition: anger, disgust, and fear.

- Geneva Vocal Emotion Stimulus Set (GVEESS) [14]: it contains 224 recordings of 12 speakers who acted 14 emotions. The recordings were sampled at 44.1 kHz. In this paper four of the 14 emotions of the database are considered for the automatic recognition: anger, disgust, fear, and desperation.
- enterface05 database [4]: this database contains 1317 audio-visual recordings with 6 emotions produced by 42 speakers. In this paper three of the six emotions of the database are considered for the automatic recognition: anger, disgust, and fear. In this database, each subject was instructed to listen to six successive short stories. After each story the subject had to react to the situation by speaking predefined phrases that fit the short story.

3.3 Results

Tables 1 and 2 contain the accuracies obtained with the features extracted from voiced and unvoiced segments, respectively. The number of Gaussian components is also indicated in the tables. In Table 1 the highest accuracies are obtained with the LZC. Table 2 shows that the best results are reached using the combination of all features, specially in GVEESS, and enterface05 databases. Note also that in this case it is possible to achieve similar results than in the case of features estimated only for voiced segments, or features estimated without the segmentation process as in related works [2, 7].

Table 1 Emotion recognition accuracies for features estimated from voiced segments

Features	GVEESS		Berlin		enterface05	
	Accuracy	M	Accuracy	M	Accuracy	M
DC	57.1 ± 14.6	4	62.7 ± 13.9	5	47.6 ± 3.8	4
LLE	68.0 ± 16.2	4	67.6 ± 8.1	5	52.1 ± 4.9	6
HE	68.1 ± 28.0	4	67.6 ± 8.1	5	52.0 ± 4.9	6
LZC	$\mathbf{82.0 \pm 11.3}$	3	$\mathbf{78.3 \pm 9.9}$	3	$\mathbf{54.0 \pm 7.3}$	4
All	65.0 ± 21.2	5	79.0 ± 10.0	4	51.1 ± 8.0	5

Table 2 Emotion recognition accuracies for features estimated from unvoiced segments

Features	GVEESS		Berlin		enterface05	
	Accuracy	M	Accuracy	M	Accuracy	M
Log energy	93.4 ± 9.8	4	64.7 ± 11.1	4	46.9 ± 4.4	5
Log energy TEO	93.1 ± 8.8	6	60.8 ± 10.3	5	54.2 ± 4.9	5
SE	93.4 ± 9.8	4	71.0 ± 12.7	4	53.7 ± 5.8	4
LEE	92.3 ± 10.3	5	$\mathbf{77.2 \pm 10.9}$	4	57.0 ± 4.1	6
All	$\mathbf{99.0 \pm 2.5}$	6	69.13 ± 16.0	6	$\mathbf{63.1 \pm 15.7}$	3

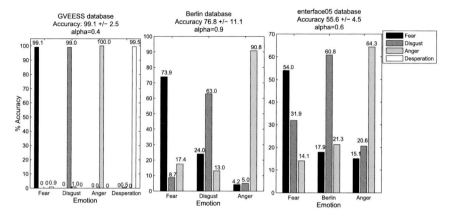

Fig. 2 Results combining the posterior probabilities

   The posterior probabilities obtained with the voiced and unvoiced features that reached highest accuracies are combined. The results are shown in Fig. 2. Note that the combination of posterior probabilities provides higher accuracy than the separated classification. In Fig. 2 each recognized emotion is indicated in the horizontal axis. The bars indicate the accuracies obtained on each emotion, i.e. in the Berlin database the bars on Disgust indicate that 24 % of the recordings labeled as Disgust where wrongly recognized by the system as Fear (bar in black). Also 13 % of the Disgust recordings where wrongly recognized as Anger. Finally, 63 % of the Disgust recordings were correctly recognized by the system as Disgust. Note that the highest accuracy in all of the three databases is obtained with the recordings labeled as Anger. This result shows that the proposed feature set is able to model the fast air flow in vocal tract produced by anger in speech, which causes vortices located near the false vocal folds providing additional excitation signals other than the pitch [19].

4   Conclusion

A total of four NLD features obtained from the WPT are extracted from speech signals to perform the automatic recognition of fear-type emotions. The voiced and unvoiced segments of each recording are characterized separately. The results show that LZC evaluated from wavelet decomposition in voice segments provides a good representation of emotional content in speech signal relative to the other NLD measures estimated from WPT. We found also that features derived from energy and entropy content of unvoiced segments are suitable for the characterization of emotional speech. The obtained results are similar to those obtained in related works when classical features are used, indicating that features related to NLD are useful to represent the emotional content in speech, and must be used to characterization of emotional speech. Additionally, the evaluation of the proposed features in speech recordings with non-controlled scenarios such as phone channels, signals with background noise, and non-acted emotions needs to be addressed in future work.

Acknowledgments Juan Camilo Vásquez-Correa is granted by the program of young researchers and innovators 2014, financed by COLCIENCIAS. Juan Rafael Orozco Arroyave is under grants of Convocatoria 528 para estudios de doctorado en Colombia 2011? financed by COLCIENCIAS. The authors thank CODI, estrategia de sostenibilidad 2014–2015? from Universidad de Antioquia.

References

1. Schuller, B., Batliner, A., Steidl, S., Seppi, D.: Recognising realistic emotions and affect in speech: state of the art and lessons learnt from the first challenge. Speech Commun. **53**(9–10), 1062–1087 (2011)
2. Li, L., Zhao, Y., Jiang, D., Zhang, Y., Wang, F., Gonzalez, I., Valentin, E., Sahli, H.: Hybrid deep neural network-hidden markov model (dnn-hmm) based speech emotion recognition. In: Proceedings of the 2013 Humaine Association Conference on Affective Computing and Intelligent Interaction, pp. 312–317 (2013)
3. Burkhardt, F., Paeschke, A., Rolfes, M., Sendlmeier, W., Weiss, B.: A database of German emotional speech. Proc. INTERSPEECH **2005**, 1517–1520 (2005)
4. Martin, O., Kotsia, I., Macq, B., Pitas, I.: The enterface'05 audio-visual emotion database. In: Proceedings of the 22nd International Conference on Data Engineering Workshops. ICDEW'06, pp. 8–15 (2006)
5. Zheng, W., Xin, M., Wang, X., Wang, B.: A novel speech emotion recognition method via incomplete sparse least square regression. Signal Process. Lett., IEEE **21**(5), 569–572 (2014)
6. Steidl, S.: Automatic Classification of Emotion-Related User States in Spontaneous Children's Speech (2009)
7. Henríquez, P., Alonso, J.B., Ferrer, M.A., Travieso, C.M., Orozco-Arroyave, J.R.: Nonlinear dynamics characterization of emotional speech. Neurocomputing **132**, 126–135 (2014)
8. Liberman, M., Davis, K., Grossman, M., Martey, N., Bell, J.: Emotional prosody speech and transcripts ldc2002s28 (2002)
9. Staroniewicz, P., Majewski, W.: Polish emotional speech database-recording and preliminary validation. In: COST 2102 Conference (Prague). Volume 5641 of Lecture Notes in Computer Science, pp. 42–49, Springer (2008)
10. Zao, L., Cavalcante, D., Coelho, R.: Time-frequency feature and ams-gmm mask for acoustic emotion classification. Signal Process. Lett., IEEE **21**(5), 620–624 (2014)
11. Hansen, J.H.L., Bou-Ghazale, S.E.: Getting started with susas: a speech under simulated and actual stress database. In: EUROSPEECH, ISCA (1997)
12. Bt Johari, N., Hariharan, M., Saidatul, A., Yaacob, S.: Multistyle classification of speech under stress using wavelet packet energy and entropy features. In: 2011 IEEE Conference on Sustainable Utilization and Development in Engineering and Technology (STUDENT), pp. 74–78, Oct 2011
13. Vasquez-Correa, J., Garcia, N., Vargas-Bonilla, J., Orozco-Arroyave, J., Arias-Londoño, J., Quintero, O.: Evaluation of wavelet measures on automatic detection of emotion in noisy and telephony speech signals. In: 2014 International Carnahan Conference on Security Technology (ICCST), pp. 1–6, Oct 2014
14. Banse, R., Scherer, K.: Acoustic profiles in vocal emotion expression. J. Person. Soc. Psychol. **70**, 614–636 (1996)
15. Boersma, P., Weenik, D.: Praat: a system for doing phonetics by computer. Report of the institute of phonetic sciences of the university of amsterdam (1996)

16. Kandali, A.B., Routray, A., Basu, T.K.: Vocal emotion recognition in five native languages of Assam using new wavelet features. Int. J. Speech Technol. **12**(1), 1–13 (2009)
17. Reynolds, D.A., Quatieri, T.F., Dunn, R.B.: Speaker verification using adapted gaussian mixture models. Digital Signal Process. **10**(1–3), 19–41 (2000)
18. Nakagawa, S., Wang, L., Ohtsuka, S.: Speaker identification and verification by combining mfcc and phase information. IEEE Trans. Audio Speech Lang. Process. **20**(4), 1085–1095 (2012)
19. Ververidis, D., Kotropoulos, C.: Emotional speech recognition: resources, features, and methods. Speech Commun. **48**(9), 1162–1181 (2006)

Assessing a Set of Glottal Features from Vocal Fold Biomechanics

Carlos Lázaro-Carrascosa and Pedro Gómez-Vilda

Abstract This paper summarizes a statistical study of a set of glottal features with the ultimate aim of measuring their capacity to discriminate and detect vocal pathology. The study is concentrated in the analysis of relevance of a set of features obtained from the analysis of phonated speech, specifically an open vowel as /ah/. The speech signal was inversely filtered to obtain the glottal source, which on its turn was used to generate a set of 72 features, describing its biometrical and biomechanical properties, among others. The study of relevance is based on factorial analysis, parametrical and non-parametrical hypothesis tests and effect size analysis, with the aim of assessing the pathologic/normophonic condition of the speaker. The validation of the results is based on discriminant analysis. The conclusions allow establishing the most relevant features and feature families for pathological voice detection. High classification rates are obtained in many cases.

Keywords Glottal features · Vocal disease · Diagnostic support · Discriminant analysis · Effect size

1 Introduction

Voice pathologies have become recently a social problem that has reached a certain concern. Pollution in cities, smoking habits, air conditioning, etc. contributes to it. This problem is more relevant for professionals who use their voice frequently: speakers, singers, teachers, actors, telemarketers, etc. Therefore techniques that are capable of drawing conclusions from a sample of recorded voice are of particular interest for the diagnosis as complementary to other invasive ones, involving

C. Lázaro-Carrascosa (✉)
Universidad Rey Juan Carlos, Móstoles, Spain
e-mail: carlos.lazaro@urjc.es

P. Gómez-Vilda
Universidad Politécnica de Madrid, Madrid, Spain
e-mail: pedro@fi.upm.es

© Springer International Publishing Switzerland 2016   209
A. Esposito et al. (eds.), *Recent Advances in Nonlinear Speech Processing*,
Smart Innovation, Systems and Technologies 48,
DOI 10.1007/978-3-319-28109-4_21

exploration by laryngoscopes, fiberscopes or video endoscopes, which are techniques much less comfortable for patients. Voice quality analysis has come a long way in a relatively short period of time. Regarding the diagnosis of diseases, we have gone in the last 15 years from working primarily with features extracted from the voice signal (both in time and frequency domains) and with scales as GRBAS [1] drawn from subjective assessment by experts, to produce more accurate evaluations with estimates derived from the glottal source. The importance of using the glottal source resides broadly in that this signal is linked to the state of the speaker's laryngeal structure. Unlike the voice signal (phonated speech) the glottal source, if conveniently reconstructed using adaptive lattices, may be less influenced by the vocal tract [2]. As it is well known the vocal tract is related to the articulation of the spoken message and its influence complicates the process of voice pathology detection, unlike when using the reconstructed glottal source, where vocal tract influence has been almost completely removed

The estimates of the glottal source have been obtained through inverse filtering techniques developed by our research group, based on the original work of Alku [3, 4]. We have also deepened into the nature of the glottal signal, dissecting and relating it to the biomechanical features of the vocal folds, obtaining several estimates of items such as mass or elasticity of vocal fold body and cover, among others. We can see the full list of features in Table 1.

In short, we have now a considerable amount of glottal features in multidimensional statistical basis, designed to be able to discriminate people with pathologic or dysphonic voices from those who do not show pathology. Four statistical techniques will be used to reach this goal: mean difference test, both parametric and nonparametric [5, 6], effect size analysis [7], factorial analysis and discriminant analysis [5, 8].

To perform the experiments we have used a balanced and robust database, consisting of two hundred speakers, one hundred of them males and one hundred females. We have also used a well-balanced proportion where subjects with vocal pathology as well as subjects who do not have a vocal pathology are equally represented.

The different statistical analyses performed will allow us to determine what features contribute in a more decisive way in the detection of vocal pathology. The issue of feature selection will be left for future studies. Therefore, some of the analyses will even allow us to present a ranking of the features based on their importance for the detection of vocal pathology. On the other hand, we will also conclude that it is sometimes desirable to perform a dimensionality reduction in order to improve the detection rates. Finally, detection rates themselves are perhaps the most important conclusion of the work, so this article will be mainly focused on them.

All the analyses presented in this work have been performed for each of the two genders in agreement with previous studies [9] showing that male and female genders should be treated independently, due to the observed functional differences between them.

Table 1 Larynx neuro-motor activity features

F01. Median of fundamental frequency f0
F02. Median of Jitter (relative between neighbor phonation cycles)
F03. Median of Shimmer (relative between mean amplitude of neighbor phonation cycles)
F04. Median of Maximum Flow Declination Rate (MFDR)
F05. Median of noise-harmonic ratio
F06. Median of Mucosal Wave Correlate to Average Acoustic Wave ratio (MWC/AAC)
F07-20. Medians of cepstral coefficients across the analysis window
F21. Median of Mucosal Wave Correlate Power Spectral Density (MWCPSD) maximum (dB)
F22. Median of MWCPSD 1st minimum (dB) rel. to F59
F23. Median of MWCPSD 2nd maximum (dB) rel. to F59
F24. Median of MWCPSD 2nd minimum (dB) rel. to F59
F25. Median of MWCPSD 3rd maximum (dB) rel. to F59
F26. Median of MWCPSD end value (dB) rel. to F59
F27. Median of MWCPSD 1st maximum position in frequency
F28. Median of MWCPSD 1st minimum position in frequency rel. to F65
F29. Median of MWCPSD 2nd maximum position in frequency rel. to F65
F30. Median of MWCPSD 2nd minimum position in frequency rel. to F65
F31. Median of MWCPSD 3rd maximum position in frequency rel. to F65
F32. Median of MWCPSD end value position in frequency rel. to F65
F33. Median of MWCPSD 1st minimum slenderness
F34. Median of MWCPSD 2nd minimum slenderness
F35-37. Medians of vocal fold body dynamic mass, losses and stiffness
F38-40. Medians of vocal fold body dynamic mass, losses and stiffness unbalances
F41-43. Medians of vocal fold cover dynamic mass, losses and stiffness
F44-46. Medians of vocal fold cover dynamic mass, losses and stiffness unbalances
F47-48. Medians of glottal source recovery instants 1 and 2
F49-50. Medians of glottal source open instants 1 and 2
F51. Median of glottal source maximum instant
F52-53. Medians of glottal source recovery amplitudes 1 and 2
F54-55. Medians of glottal source open amplitudes 1 and 2
F56-57. Median of glottal flow stop and start instants
F58. Median of glottal flow closing instant
F59-62. Medians of glottal flow gap, contact, adduction and permanent defects
F63-65. Medians of the 1st, 2nd and 3rd order cyclic coefficients (tremor)
F66-67. Medians of the physiological band tremor frequency and amplitude
F68-69. Medians of the neurological band tremor frequency and amplitude
F70-71. Medians of the flutter band tremor frequency and amplitude
F72. Median of the root mean square tremor amplitude

2   Materials and Methods

The database used for the study is a corpus of 200 subjects (100 male and 100 female), created by our research group, which evenly distributes the presence and absence of vocal pathology. Thus, the database can be described as: 50 non-pathological female subjects, 50 pathological female subjects, 50 non-pathological male subjects and 50 male pathological subjects. This data will be split in two sets for Discriminant Analysis: train and test. We have tried several sizes for the sets, in different experiments, but the more frequently proportion used is 90% for training—10% for testing.

The majority of studies analyze data collected from male and female subjects independently. In general, the need for male and female samples to reach general conclusions is well known [9]; the research in this field also recommends taking into consideration the different physiological characteristics of both genders. The ages of the subjects range from 19 to 56, with a mean of 30.02 years and a standard deviation of 9.94 years. Normal condition has been determined by electroglottography, video-endoscopy and GRBAS tests. Besides, previously some criteria which patients must meet to fulfill such a condition were established:

- self-report not having any laryngeal pathology;
- have a voice according to gender, age and cultural group speaker, plus a suitable pitch, timbre, volume and flexibility of diction;
- be a non-smoker;
- no history of any surgery related to any laryngeal pathology;
- no history of endotracheal intubation in the last year.

Concerning pathological cases, the sample set contains around the same number of mild (functional) and moderate pathology grades, comprising polyps, nodules and Reinke's edemae. The recording protocol comprises three different utterances of the vowel /ah/ with duration longer than 3 s, taken with a headset microphone placed 10 cm from the speaker's mouth. The vowel "ah" has been selected because it is easy to pronounce (causes less tension), and, above all, because the vocal tract influence is better removed when we use this vowel, according to various tests performed. The signals were sampled at 44,100 Hz, lately undersampled to 22,050 Hz. The resolution used was 16 bits. The recordings were performed using the external sound board Creative Studio. Segments of 0.2 s long were extracted from the central part (the more stable) for analysis and featureization. These segments included a variable number of cycles of phonation: about twenty in male voices, with a fundamental frequency of 100 Hz and about forty in female voices, with a fundamental frequency of 200 Hz [10].

The protocol used in the experimentation is based on the following steps:

1. Mean difference parametric test (Student's t) between the two groups of the study: people with speech pathology and people free from speech pathology. An additional effect-size analysis was also included to assess the relevance of the differences found.

2. Nonparametric contrast between the two groups mentioned. In particular, test of Mann-Whitney. This test, as well as the one mentioned in the previous point was performed on all the features, in order to facilitate the comparison of results.
3. Factor Analysis based on Principal Components applied on the full set of features, and also on significant subsets thereof. These subsets of features come from two sources: first of them have been prepared from the results of the previous points of this protocol: we will use the features that give a significant difference between the two groups mentioned above, we will also use the features with the highest effect size values and we will finally use the full set of features except the ones with worse effect size values. On the other hand, some subsets of features correspond directly with the family to which they belong—Perturbation, Cepstrals, Biomechanics, Biometrics, Temporals, Contact or Tremor—and the last ones are formed by all the features except those belonging to one family at a time.
4. Discriminant Analysis on the full set of features and on significant subsets thereof. The subsets are prepared with the same criterions as in the previous point. We will also apply this analysis to the factors obtained in the previous section, and to the best features—on their own—obtained from the effect size analysis.

3   Results

The rates of the first studies performed can be seen in Table 2. This table offers an overall rate, calculated as the arithmetic mean between the rate for the unselected cases and the rate for a process of cross-validation that consists on classifying each case using the functions derived from all cases other than that case.

We can see in the table that the rates are better when we use factors in the female database. There, the highest rate (92.73 %) is achieved for the factors derived from the significant features in the t-student test. In the male database the difference between the use of features or factors is fewer. The highest rate (79.39 %) is achieved for the significant features in the Mann-Whitney test.

The features whose effect size value can be considered *large* by Hopkins [7] in the female database were the following: F58 (effect size value: 1.88), F56 (1.59), F51 (1.52), F21 (1.51), F25 (1.43), F42 (1.41), F57 (1.29) and F24 (1.15). It is remarkable that the first three values come from temporal features, next three are biometric or biomechanical features, and last two are temporal and biometric features again. Some of these features (F56, F51, F21, F42 and F57) achieve by themselves overall detection rates above 80 %, but we would like to highlight the feature number 58 (Median of Glottal Flow Closing Instant) that produces a detection rate of 92.16 %. We can see the boxplot corresponding to this feature in Fig. 1.

Regarding the male database, we have just found one *large* value in the effect size analysis, corresponding to the feature F50 (temporal, 1.21). Besides, the highest values, that can be considered *moderate* [7] are achieved by the features F42 (biomechanic, 1.14), F51 (temporal, 1.17) and F25 (biometric, 1.00). From all these we can

Table 2 First test performed and detection rates obtained for the bases of male and female data

Test	Female data	Male data
	Avg. value (%)	Avg. value (%)
All the features	78.40	78.89
All the factors	87.79	75.17
T-student positive features	83.00	77.22
T-student positive factors	92.73	76.44
Mann-Whitney positive features	88.50	79.39
Mann-Whitney positive factors	90.61	79.17
Effect size best features	85.46	74.44
Effect size best factors	87.02	79.06
Effect size without worst features	85.73	76.28
Effect size without worst factors	89.44	76.66

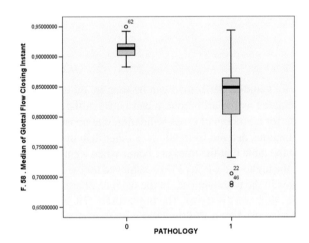

Fig. 1 Boxplot of the feature 58, female database

remark the feature F51 (Median of Glottal Source Maximum Instant), that achieves an overall detection rate of 78.61 %.

The results related to the differerent families of features for the female database can be seen in Figs. 2 and 3. Figure 2 is divided in two sections: the left one is referred to the features of each family; the right one is referred to the factors generated with the features of each family. Figure 3 is also divided in two: the left section is referred to all the features except the ones included in each family; the right section is referred to the factors generated with all the features except the ones included in each family.

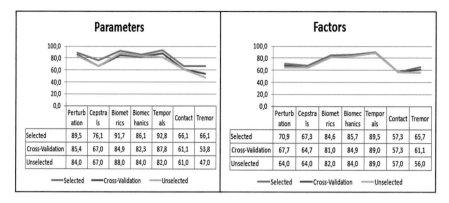

Fig. 2 Study related to the feature families, female database, part one

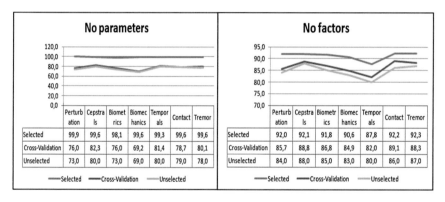

Fig. 3 Study related to the feature families, female database, part two

There are four families whose features offers quite high rates by themselves: Perturbation, Biometrics, Biomechanics and Temporals. The last two of these families improves their results when factors are applied. According to these results, if we remove the Cepstral, Contact or Tremor features we obtain high success rates, very specially when we use factors.

The results related to the male database can be seen in Fig. 4. We can highlight the behaviour of the Temporal and Biomechanic features this time. They are the families that produce higher rates. Besides, it is remarkable that there are no families with very low rates in the male database. In general, the use of factors procures worse results. On the other hand, when we remove the different families we obtain the best rates, improving the results obtained in the previous experiments: several overall rates are above 80 %, and the highest one correspond to the remove of the biomechanics features: 81.11 % of success.

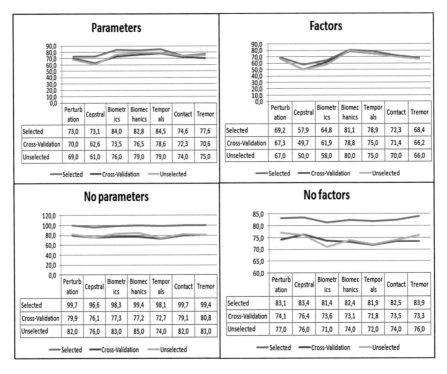

Fig. 4 Study related to the feature families, male database

4 Conclusions

We have carried out a statistical study of glottal features using a database composed of individuals of both genders. The study has mainly focused on finding out the most suitable features to discriminate pathological and non-pathological voices, using the Discriminant Analysis technique. From the results shown we reach the following conclusions:

- We obtain better rates with female data than with male data in all cases.
- Contact, Tremor and Cepstral features seem to have less discriminatory power than other categories, especially in the female database.
- In general, factors work better with female data; features do it with male data.
- We obtain good results when we use a large number of features (or factors). The use of all features (or factors) with or without the removing of some certain features reaches generally one of the best results in both databases.
- Not contradicting the previous point, the categories of features considered are able to obtain high classification rates by themselves, just like the best features according to the effect size analysis, which produce even better results in some cases.

- The best results are obtained with the significant factors from the t-student test for the female database (92.73 %) and with all the features except the biomechanical ones for the male database (81.11 %).

Acknowledgments This work is being funded by grants TEC2012-38630-C04-01 and TEC2012-38630-C04-04 from Plan Nacional de I+D+i, Ministry of Economic Affairs and Competitiveness of Spain

References

1. Hirano, M., et al.: Acoustic analysis of pathological voice: some results of clinical application. Acta Otolaryngologica **105**(5–6), 432–438 (1988)
2. Gómez, P., et al.: Glottal Source biometrical signature for voice pathology detection. Speech Commun. **51**, 759–781 (2009)
3. Alku, P.: An automatic method to estimate the time-based parameters of the glottal pulseform. In: ICASSP'92, pp. II/29–32 (1992)
4. Alku, P.: Parameterisation methods of the glottal flow estimated by inverse filtering. In: Proceedings of VOQUAL'03, pp. 81–87 (2003)
5. Härdle, W., Simar, L.: Applied Multivariate Statistical Analysis, vol. 22007. Springer, Berlin (2007)
6. Ho, R.: Handbook of Univariate and Multivariate Data Analysis and Interpretation with SPSS. CRC Press, Florida (2006)
7. Hopkins, W.G.: A new view of statistics. http://www.sportsci.org/resource/stats/index.html (2000)
8. Klecka, W.R. (ed.): Discriminant Analysis, vol. 19. Sage, USA (1980)
9. Ruiz, M.T., Verbrugge, L.M.: A two way view of gender bias in medicine. J. Epidemiol. Community Health **51**, 106–109 (1997)
10. Gómez, P., et al.: PCA of perturbation parameters in voice pathology detection. In: Actas de INTERSPEECH-2005, pp. 645–648 (2005)

Pitch Estimation Based on the Cepstrum Analysis by the Multi Scale Product of Clean and Noisy Speech

Wided Jlassi, Aicha Bouzid and Noureddine Ellouze

Abstract In this paper we propose a new method for estimating the pitch from the speech signal which consists of analysing real cepstrum by the multiscale product (MP) using continuous wavelet transform (WTC) having one vanishing moment (CAMP). Our approach to estimate the pitch consists of the following steps: first we frame the voiced signal, second we calculate the real cepstrum of each frame. Finally, we compute the MP of the cepstrum. The MP is the product of the WTC at three scales. Our method will be evaluated by the Keele database under clean and noisy conditions. Experimental results indicate that the gross pitch errors (GPE) are lower than the compared methods under clean and noisy conditions.

Keywords Speech · Real cepstrum · Wavelet transform · Multi-scale product · Pitch

1 Introduction

The fundamental frequency of the speech signal is the frequency of vibration cycle of the vocal cords, determined by the tension of the muscles that control them. The fundamental frequency varies from one speaker to another according to age and sex [1]. The estimation of F0 is very important in the field of speech processing for many applications such as coding [2], analysis or speech recognition [3].

Various methods in the frequency, time and time-frequency domain have been proposed for pitch estimation [4] proposes a comparative overview.

W. Jlassi (✉) · A. Bouzid · N. Ellouze
Electrical Engineering Department National School of Engineers of Tunis
Le Belvédère, BP. 37, 1002 Tunis, Tunisia
e-mail: j_wided@yahoo.com

A. Bouzid
e-mail: bouzidacha@yahoo.fr

N. Ellouze
e-mail: n.ellouze@enit.rnu.tn

© Springer International Publishing Switzerland 2016
A. Esposito et al. (eds.), *Recent Advances in Nonlinear Speech Processing*,
Smart Innovation, Systems and Technologies 48,
DOI 10.1007/978-3-319-28109-4_22

The present article introduces a new method for F0 estimation based on the cepstrum analysis by the MP that operates in the time-scale domain.

Our algorithm operates the MP of the voiced speech cepstrum using a wavelet having one vanishing moment under clean and noisy conditions.

Besides, our approach is evaluated and compared to the Cepstrum and Praat approaches [5]. Praat is a free software designed for processing and synthesis of speech sounds (phonetic). It determines the F0 candidates by the location of maxima in the cross-correlation function over the voiced speech.

This paper is organized as follows: Sect. 2 describe the classic cepstrum method for the pitch estimation. Section 3 presents some properties of the continuous wavelet transform and the Multi-scale Product for the detection of signal singularities. In Sect. 4, the details of our approach for estimating the pitch are exposed. In Sect. 5, our approach is evaluated using the Keele University database under clean and noisy conditions. Finally, we conclude this work.

2   Cepstrum

According to the speech production model, the voiced speech is the convolution of the glottal excitation by the vocal tract transfer [6].

$$s(n) = e(n) * \theta(n) \tag{1}$$

where $s(n)$ is the speech signal, $e(n)$ is the excitation signal and $\theta(n)$ is the contribution of the vocal tract.

The cepstral analysis consists in ensuring the deconvolution [7].

The real cepstrum is defined as:

$$c[n] = \frac{1}{2\pi} \int_{-\pi}^{\pi} \log |S(w)| e^{jnw} dw \tag{2}$$

where:

$$S(w) = \sum_{n=-\infty}^{+\infty} s(n) e^{-jwn} \tag{3}$$

A flowchart of the cepstrum analysis is shown in Fig. 1.

To get an estimation about the fundamental period and the fundamental frequency we look for a peak in the appropriate quefrency region:

$$F0 = 1/\text{quefrency} \tag{4}$$

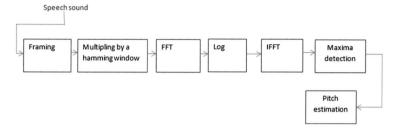

Fig. 1 Flowchart of the cepstrum analysis

3 Wavelet Transform and Multi Scale Product

The WT is a well-known method for the detection of signal singularities such as the glottal closure instant GCI in the speech signal [8].

In fact, discontinuity in the signal is detected by a wavelet transform using a wavelet having one vanishing moments [9].

The wavelet transform is expressed as the nth derivative of the signal smoothed by a function $\bar{\theta}$

$$\mathrm{Wf}(\mathrm{u, s}) = \mathrm{s}^{\mathrm{n}} \frac{\mathrm{d}^{\mathrm{n}}}{\mathrm{du}^{\mathrm{n}}} (\mathrm{f} * \bar{\theta}_{\mathrm{s}})(\mathrm{u}) \qquad (5)$$

where

$$\bar{\theta}_s(t) = \frac{1}{\sqrt{s}} \theta\left(\frac{-t}{s}\right) \qquad (6)$$

A wavelet has n vanishing moments if and only if:

$$\int_{-\infty}^{+\infty} t^k \Psi(t) dt = 0 \qquad (7)$$

With $\forall\ 0 \leq k \leq n-1$ and n indicates the number of the vanishing moments characterizing the wavelet.

Bouzid et al. [8] has proved that the Multi-scale Product (MP) improve the singularity detection on the speech signal

$$p(n) = \prod_{i=1}^{3} W_{s_i} f(n) \qquad (8)$$

where $W_{s_i} f(n)$ is the WT of the function f at the scale Si.

4 The Proposed Method

In this section, we present the CAMP approach for the estimation of the fundamental frequency.

The first step of the algorithm consists in framing the voiced signal and multiplying each frame by a Hamming window. Thereafter, we proceed to calculate the real cepstrum for each windowed frame and the WT of the cepstrum at three scales. The next step is operates the product of WT coefficient and detect the local zero crossing corresponding to the pitch period of each analyzed frame. Finally, we calculate the corresponding fundamental frequency before proceeding to a post-processing phase.

This approach is applied using the first derivative of a Gaussian as a wavelet having one vanishing moment.

The pitch period corresponds to the position of the zero crossing between the maximum and the minimum detected on the MP of the cepstrum.

The focus of the post-processing is to reduce the F_0 estimation error under both clean and noisy conditions in order to have more reliable pitch detection algorithm.

This smoothing phase consists in comparing each value of the F_0 by its previous and next one and if there is a difference bigger than 30 Hz, we consider that is a false measure and is replaced by the average of the last two ones.

Figures 2 and 3 show the speech signal, followed by its cepstrum, and the CAMP for the vowel /e/ pronounced by the speaker f1 and the same vowel corrupted by the white noise (−5 dB) respectively.

We note that the peaks given by the proposed approach is revealed clearer and the spurious peaks in the cepstrum are almost eliminated even if the speech signal is noisy.

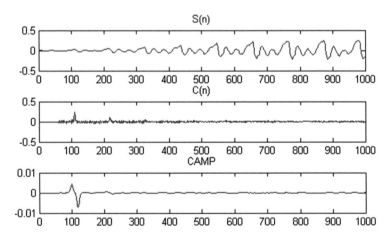

Fig. 2 The speech signal, the cepstrum and the CAMP for the vowel /e/ pronounced by the speaker f1

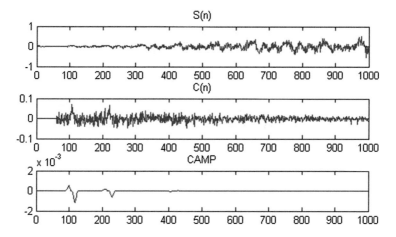

Fig. 3 the speech signal, the cepstrum and the CAMP for the vowel /e/ pronounced by the speaker f1 corrupted by a white noise with SNR of −5 dB

5 Evaluation and Comparison

The Keele speech database provided by the Keele University [10] is used to evaluate the performance of our method. It includes ten speakers, five men and five women, pronouncing the text "The North Wind Story"

The sampling frequency used in this database is 20 kHz.

The parameters of our proposed approach are as follows:

The three scales (s1, s2, s3) used are (1, 1.5, 2) for women and (1.5, 2.5, 3.5) for men.

The Hamming window has a length of 2048 samples for men and 1024 samples for women, whereas the frame shift is fixed to 200 samples for both. These lengths allow a good compromise between efficiency and stability of the parameters to be measured.

The fundamental frequency references are obtained by the autocorrelation function of the multiscale product of the larygograph signal. This choice is justified by the fact that it has been proven that the MP of the EGG signal is the most effective to measure the fundamental frequency [11].

To evaluate the performance of our proposed method, the gross pitch error (GPE) is considered if

$$\left| \frac{f0_{i,\text{estimated}}}{f0_{i,\text{reference}}} \right| > \epsilon \tag{9}$$

And the GPE rate is defined as:

$$\text{GPE} = \frac{N_{\text{GE}}}{N_{\text{V}}} \times 100\% \tag{10}$$

Table 1 GPE(%) of Praat, cepstrum and the CAMP methods

Methods	Clean	White noise			
		−5 dB	0 dB	5 dB	10 dB
Praat	3.22	30.91	11.53	6.11	4.28
Cepstre	2.97	31.41	20.70	10.91	6.46
CAMP	2.23	24.57	12.89	4.63	2.75

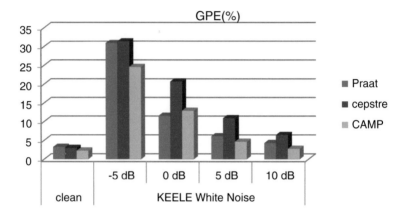

Fig. 4 GPE(%) of the Praat, cepstrum and CAMP methods using Keele database

where N_V the numbers of frames supposed voiced.

i is defined as the frame index and is the threshold which is usually taken 20 %.

Table 1 reports evaluation results for gross pitch error (GPE) of the proposed method under clean and noisy conditions.

Figrue 4 shows that our CAMP method gives very interesting results compared to Praat and Cepsrtrum under clean and noisy conditions with the lowest GPE rate using a wavelet having one vanishing moments.

Tables 1 and 2 shows the execution time (in second) of our method compared to the cepstrum method.

Referring to the results of Table 2, time of execution of the cepstrum method is better than our method.

Table 2 Execution time (in second) of our method compared to the cepstrum method

Methods	Clean	White noise			
		−5 dB	0 dB	5 dB	10 dB
Cepstre	17.41	11.75	13.11	12.29	12.74
CAMP	25.47	30.24	26.27	30.35	25.64

6 Conclusion

In this paper, we propose a new method for estimating the pitch period from the speech signal based on the cepstrum analysis by the multi-scale product under clean and noisy condition.

Our method can be summarized in three essential steps. First, we frame the voiced signal and multiply each frame by a Hamming window. Second, we calculate the real and its WT at three scales. Thus we calculate the product obtained by the WT and detect the local zero crossing corresponding to the pitch period. Finally, we calculate the corresponding fundamental frequency before proceeding to a post-processing step.

This approach significantly improves the cepstrum by eliminating the most uninteresting peaks in clear and noisy conditions.

Our approach is evaluated and compared to the Cepstrum and Praat approaches. It gives the best results in terms of gross pitch errors.

Future work concerns, the evaluation of our approach in different noisy environments.

References

1. Wang, M., Lin, M.: An Analysis of Pitch in Chinese Spontaneous Speech. In: International Symposium on Tonal Aspects of Tone Languages, Beijing, China (2004)
2. Spanias, A.S.: Speech coding: a tutorial review. Proc. IEEE **82**, 1541–1582 (2004)
3. De Bot, K.: Visual feedback of intonation I: effectiveness and induced practice behavior. Lang. Speech **26**, 331–350 (1983)
4. Hess, W.: Pitch Determination of Speech Signals: Algorithms and Devices. Springer, Berlin and Heidelberg (1983)
5. Boersma, P.: Praat, a system for doing phonetics by computer. Glot. Int. **5**(9/10), 341–345 (2001)
6. Noll, A.M.: Cepstrum pitch determination. Acoust. Soc. Am. **41**, 293–309 (1967)
7. Makhljanl, R., Shrawankar Hrawankar, U., Thakare, V.M.: Speech enhancement using pitch detection approach for noisy environement. Int. J. Eng. Sci. Technol. (IJEST) **3**(2), (2011)
8. Bouzid, A., Ellouze, N.: Electroglottographic measures based on GCI and GOI detection using multiscale product. Int. J. Comput. Commun. Control **3**(1), 21–32 (2008)
9. Mallat, S.: A Wavelet Tour of Signal Processing, 2nd edn. Academic Press (1999); Detection of speech signals. IEEE Trans. Inf. Theory **38**, 917–924 (1992)
10. Plante, F., Meyer, G., Ainsworth, W.A.: A pitch extraction reference database. In: EUROSPEECH, pp. 837–840 (1995)
11. Bouzid, A., Ellouze, N.: Voice source measurement based on multiscale analysis of electroglottographic signal. Speech Commun. (2009)

Minimizing Free Energy of Stochastic Functions of Markov Chains

Rita Singh

Abstract Automatic speech recognition has generally been treated as a problem of Bayesian classification. This is suboptimal when the distributions of the training data do not match those of the test data to be recognized. In this paper we propose an alternate analogous classification paradigm, in which classes are modeled by thermodynamic systems, and classification is performed through a *minimum energy* rule. Bayesian classification is shown to be a specific instance of this paradigm when the temperature of the systems is unity. Classification at elevated temperatures naturally provides a mechanism for dealing with statistical variations between test and training data.

Keywords Bayes classification · Free energy · Temperature · Speech recognition

1 Introduction

In the usual rule for Bayesian classification any data X is assigned to the class that is most likely to have generated it. Formally, if we represent the class assignment of X as $c(X)$, the classification rule is given by

$$c(X) = \arg \max_{C} P(C) P(X|C) \tag{1}$$

where C represents any class, $P(C)$ is the a priori probability of C, and $P(X|C)$ is the probability distribution of data from class C. In the context of automatic speech recognition, the classes are actually word sequences [1]. The Bayes classification rule attempts to identify the a posteriori most likely word sequence, given (features derived from) a recording X.

The Bayes classification rule is optimal when $P(X|C)$ is the *true* class-conditional probability distribution of data for C. In practical scenarios, however, the true

R. Singh (✉)
Carnegie Mellon University, Pittsburgh, PA 15213, USA
e-mail: rsingh@cs.cmu.edu
URL: http://cs.cmu.edu/~rsingh

© Springer International Publishing Switzerland 2016
A. Esposito et al. (eds.), *Recent Advances in Nonlinear Speech Processing*,
Smart Innovation, Systems and Technologies 48,
DOI 10.1007/978-3-319-28109-4_23

distribution $P(X|C)$ is not known and must be approximated by a *model* $\hat{P}(X; \Lambda_C)$, the parameters Λ_C of which must be learned from training data. The model $\hat{P}(X; \Lambda_C)$ is generally learned to be close (in the sense of KL divergence) to the distribution of the training data, and often does not adequately model the *test* data. As a result, the classification rule of (1) is suboptimal and can result in significantly degraded performance when used in a speech recognizer.

In [2, 3] we have proposed an alternative formalism for classification in such scenarios. Instead of assuming that $\hat{P}(X; \Lambda_C)$ represents a class-specific probability distribution, we interpret it as a *thermodynamic system*, which has resulted in an observation X. Subsequently, we replace the "maximum probability" criterion deriving from the stochastic-process interpretation which leads to Bayes classification rule, with a "minimum energy" criterion: the observation X is now assigned to the class C whose system must expend the least energy to generate it. Designating the energy as $F_C(X)$, the modified classification rule is

$$c(X) = \arg\min_C F_C(X) \tag{2}$$

The distinction between the probability-based rule of (1), and (2) can be reconciled by defining $F_C(X) = -\log P(C, X; \Lambda_C)$. Indeed, such an equivalence is commonly ascribed, and has been drawn upon in the definition of stochastic models such as the Gibbs distribution [4], or even the normal distribution, where the log probability is analogous to common definitions of energy associated with a data or vector [5].

Thermodynamic systems, however, also include a *temperature* parameter. In the physical world the temperature of the system characterizes the fluctuation of state of the system, effectively characterizing the variation in any measurements of it—the greater the temperature the greater the variation will be. In our classification framework, the temperature parameter may be analogously considered as characterizing the increased variation in observations. At the specific setting of $T = 1$, the probabilistic and energy classification rules become identical; at higher values however, the energy-based mechanism naturally allows for greater variation in the data, such as the differences between training and test data.

In the subsequent sections we will first describe the general Themodynamic principle of free energy (Sect. 2), followed by a brief outline of minimum-free energy classification (Sect. 3) and how it applies to automatic speech recognition (Sect. 4). We then present experimental evidence of the effectiveness of the formulation (Sect. 5) and discussion (Sect. 6).

2   Free Energy of a Stochastic System

A thermodynamic system at temperature T can exist in one of a large (potentially infinite) number of states [6]. At each state s the system has an energy E_s. If the probability of state s is given by $P_T(s)$, the *internal energy* of the system, representing the capacity of the system to do work, is given by the average: $U_T = \sum_s P_T(s)E_s$. This capacity is counteracted by its internal disorder, which is factored into its entropy $H_T = -\sum_s P_T(s) \log P_T(s)$ and the temperature T of the system. The *Helmholtz free energy* of the system, measuring the useful work obtainable from the system when it is closed, is thus defined by

$$F_T = U_T - T H_T = \sum_s P_T(s)E_s + T \sum_s P_T(s) \log P_T(s) \qquad (3)$$

At constant temperature, systems will drift towards the lowest free-energy states [6], adjusting probabilities $P_T(s)$ until F_T is minimized. The distribution $P_T(s)$ at *thermal equilibrium*, obtained by minimizing F_T, is the Gibbs distribution

$$P_T(s) = \frac{1}{Z} \exp\left(\frac{-E_s}{T}\right) \qquad (4)$$

where Z is a normalizing term. The corresponding *equilibrium free energy* is

$$F_T = -T \log \sum_s \exp\left(\frac{-E_s}{T}\right) \qquad (5)$$

3   Classification with Free Energy

Consider a class with a stochastic generative latent-variable model that assigns a probability $P(X|C) = \sum_s P(s|C)P(X|C, s)$ to any observation. To generate any observation, the generative process must be in any latent state s and draw an observation from the state-conditioned distribution $P(X|C, s)$.

For energy-based classification, we model every class C instead by a thermodynamic system that can exist in one of a set S_C of states. Within any state s the system must have an energy $E_s^C(X)$ to result in the observation X. The equilibrium free energy of this system, when it is at temperature T, is hence given by

$$F_T^C(X) = -T \log \sum_s \exp\left(\frac{-E_s^C(X)}{T}\right) \qquad (6)$$

The "energy" of each state is equated to negative log-likelihood of the combination of the state and the observation, $E_s^C(X) = -\log P(X, s, C)$—intuitively, the greater the energy needed to exhibit X, the less likely the system is to visit the corresponding state. Using these values, the free energy of the system for any class comes out as

$$F_T^C(X) = -\log P(C) - T \log \sum_s \exp\left(\frac{\log P(X, s|C)}{T}\right) \qquad (7)$$

We specify the minimum-energy classification rule as follows: the observation X is assigned to the class that has the lowest free energy for X.

$$c(X) = \arg\min_C F_T^C(X) = \arg\min_C \left(-\log P(C) - T \log \sum_s P(X, s|C)^{\frac{1}{T}} \right) \quad (8)$$

This is a natural extension of the principle that thermodynamic systems evolve towards minimum-energy configurations. Note that the objective in (8) remains a function of the temperature parameter T. As T increases and the internal disorder in the systems increases, the systems for the various classes will more frequently visit low-energy states associated with X; in the limit, T dominates and all classes are equally capable of generating observation X. From a classification perspective, T characterizes external influences such as noise or other factors that increase the entropy of the systems. Note that at $T = 1$ (the "quiescent" condition) (8) reduces to the conventional Bayesian classifier of (1).

4   Minimum Free Energy Decoding with Hidden Markov Models

A particularly interesting family of stochastic models that can be cast into the free-energy framework are stochastic functions of Markov chains, also known as Hidden Markov Models (HMMs) [7]. HMMs are frequently employed in automatic speech recognition systems [1]. HMM-based speech recognition systems formulate the Bayes classification paradigm as identifying the word sequence with the a posteriori most likely state sequence for any speech recording X [8].

$$\hat{W} = \arg\max_W \max_S P(W)P(S|W)P(X|S, W) \quad (9)$$

where W represents any word sequence, $P(W)$ is the a priori probability of W, S is a state sequence through the HMM for W, and $P(S|W)$ represents its probability. The state output distributions of the HMM are often modeled by mixture distributions, typically Gaussian mixtures. Thus the classification equation can be re-written as

$$\hat{W} = \arg\max_W \max_S P(W, S) \prod_{t=1}^{T} P(X_t|s_t)$$

$$= \arg\max_W \max_S P(W, S) \prod_{t=1}^{T} \sum_k P(k|s_t) N(X_t; \mu_{s_t,k}, \Sigma_{s_t,k}) \quad (10)$$

Here X_t is the tth vector in X and s_t is the tth state in S. $N()$ represents a Gaussian distribution, $P(k|s_t)$ represents the mixture weight of the kth Gaussian in the Gaussian mixture distribution for state s_t, and $\mu_{s_t,k}$ and $\Sigma_{s_t,k}$ represent the mean and covariance of the kth Gaussian in s_t.

We can define $K = k_1, k_2, \ldots, k_T$, representing a sequence of Gaussians, one each from states $s_1 - s_T$. If each state is represented by a mixture of M Gaussians, there are M^T such sequences. Combining W and S into a single variable $\mathcal{W}$, (10) can be rewritten as:

$$\hat{W} = \arg\max_{\mathcal{W}} \log P(\mathcal{W}) + \log \sum_{K} P(X, K|\mathcal{W}) \tag{11}$$

where

$$P(X, K|\mathcal{W}) = \prod_{t=1}^{T} P(k_t|s_t) N(X_t; \mu_{s_t,k}, \Sigma_{s_t,k}) \tag{12}$$

The above is now easily recast into minimum-energy classification. Each class $\mathcal{W}$ is represented by a thermodynamic system, which can be in one of a M^T states, where each state is a Gaussian sequence K. The energy of any state is given by $E_K^{\mathcal{W}} = -\log P(X, K, \mathcal{W})$. Consequently, the minimum free-energy classification rule of (7) becomes, with minimal manipulation,

$$\hat{W} = \arg\min_{\mathcal{W}} -\log P(\mathcal{W}) - T \sum_{t} \log \sum_{k} P(k_t|s_t)^{\frac{1}{T}} N(X_t; \mu_{s_t,k}, \Sigma_{s_t,k})^{\frac{1}{T}} \tag{13}$$

This modified classification rule requires only *minimal* changes to the conventional Viterbi decoder. The computation of state output distribution values as $\sum_k P(k_t|s_t) N(X_t; \mu_{s_t,k}, \Sigma_{s_t,k})$ is replaced by $\left(\sum_k P(k_t|s_t)^{\frac{1}{T}} N(X_t; \mu_{s_t,k}, \Sigma_{s_t,k})^{\frac{1}{T}}\right)^{T}$. The rest of the decoder remains unchanged. We refer to this modified decoding strategy as "minimum-energy decoding".

5 Experiments

We expect the benefits of minimum-energy decoding to be exhibited primarily when there is mismatch between the distributions employed by the recognizer and the test data. Our experiments were therefore aimed at evaluating the effect of minimum-energy decoding under conditions of mismatch. One of the most common reasons for mismatch in speech recognition systems is noise: test data to be recognized will frequently be corrupted by various types of noise not seen in the training data. Note that noise robustness is not the focus of this paper; rather it is the mismatch between the acoustic models and the test data.

We conducted experiments on the Fisher database [9], digitally corrupted by noise to introduce mismatch. The training data comprised the Fisher Phase I corpus (LDC catalog No. LDC2004S13), including 5,850 two-channel audio files, each

Table 1 Performance of minimum-energy speech recognition in terms of word error rate (%)

Temp (dB)	1.0	1.1	1.2	1.3	1.4	1.5	1.6	1.7	1.8	1.9	2.0
0	92.7	90.8	87.4	85.4	79.9	79.3	78.2	77.9	**75.8**	77.8	82.3
5	65.3	62.4	57.4	55.8	53.4	51.4	50.5	49.2	**48.3**	51.3	57.7
10	47.6	46.9	45.1	44.8	42.8	38.2	37.4	**36.6**	37.8	41.2	47.2
15	36.2	36.1	35.1	33.5	31.9	**30.2**	30.8	31.8	34.2	37.2	40.1
20	27.2	26.8	25.1	**24.2**	24.6	25.4	27.2	29.4	32.4	35.2	38.1

The $T = 1.0$ column corresponds to conventional decoding, representing Bayesian classification. The bold numbers are the best results obtained in each row

containing a full conversation of up to 10 min. 111,157 speech segments from the corpus, representing nearly the entire data, minus our held-out test set, were used to train the models. A set of 10,000 segments from the same data were used as our held-out designated test set. The test set was corrupted to various signal-to-noise ratios (SNR) by babble noise to introduce mismatch with respect to the the training set.

We used the Carnegie Mellon University's Sphinx-3 triphone-based automatic speech recognition system [10] to perform all our experiments. All models were 3-state left-to-right Bakis topology HMMs. A total of 5000 tied states, each modeled by a mixture of 16 Gaussians, were employed. The language model was trained from the Fisher training corpus and the Switchboard corpus. The baseline recognition word error rate on the uncorrupted test set was 14.3 %.

The test data were recognized at several temperatures. Table 1 shows the word error rates obtained at each SNR, against the temperature at which the data were decoded. The column in the table corresponding to $T = 1.0$ is identical to the standard Bayesian decoding, as explained earlier.

We note from the results that the optimal recognition performance is *not* obtained at $T = 1$. The best result in all cases occurs at an elevated temperature. Moreover, as the SNR decreases and, consequently, the degree of mismatch between the training and test data increases, the optimal temperature increases. Thus, while the optimal temperature at 0dB is close to 2.0, at 20 dB the optimal temperature is 1.3. At greater mismatch, e.g. at 0dB, the improvement from increased temperature is quite dramatic, amounting to about 17 % absolute.

6   Conclusions

Elevation of temperature is observed to result in significantly improved recognition under conditions of mismatch. Considering that just a simple adjustment has been made to the manner in which state-output probabilities are computed during decoding in order to achieve this, the improved classification scheme is promising for use in speech recognizers. It must be noted that although these improvements are not as large as that improved with sophisticated noise compensation algorithms, that is not the objective of our solution. The proposed algorithm makes no assumptions about the *reason* for the mismatch; the only assumption is that while the systematic differences between classes persist in the test data, the actual distribution may be shifted with respect to the training data. Our purpose is to demonstrate that the proposed approach, which is a natural extension of conventional Bayesian classification, could be used to good effect under such conditions.

A key question that remains to be answered is that of *selecting* the optimal temperature in an unsupervised manner. We continue to explore this problem.

More generally, the notions of "temperature" and "free energy" have often been invoked in the context of annealing for optimization of objective functions defined over a continuous support [11]. Classification, on the other hand, is typically a search over a discrete support, and not usually viewed as an optimization problem. This is generally considered to be distinct from the situations where notions of free energy and temperature may be invoked. The novelty of our approach is to view the latter as a special case of optimization, where the task is to find the optimal value over a discrete support. In this context, automatic speech recognition systems present an interesting case—although the support remains discrete, it is inifinite, representing all possible sentences that may be spoken, suggesting that the concept of annealing may be drawn upon if the search space could somehow be ordered and represented over a continuum. However, how this may be done is unclear, and this remains a topic for future research.

References

1. Singh, R., Raj, B., Virtanen, T.: The basics of automatic speech recognition. In: Techniques for Noise Robustness in Automatic Speech Recognition. John Wiley and Sons Inc. (2012)
2. Singh, R.: Audio classification with thermodynamic criteria. In: Proceedings IEEE International Workshop on Cloud Computing for Signal Processing, Coding and Networking (2014)
3. Singh, R., Kumatani, K.: Free energy for speech recognition. In: Proceedings International Conference Acoustics, Speech and Signal Processing (2015)
4. Landau, L.D., Lifshitz, E.M.: Statistical Physics: Course of Theoretical Physics, vol. 5, 3rd edn, p. 12. Pergamon Press, Oxford (1980)
5. Ranzato, M.A., Boureau, Y.L., Yann, L.C.: Sparse feature learning for deep belief networks. Proc. Adv. Neural Inf. Process. Syst. **21**, 1185–1192 (2008)
6. Callen, H.B.: Thermodynamics and an Introduction to Thermostatistics. John Wiley and Sons Inc. (1985)
7. Baum, L.E., Petrie, T.: Statistical inference for probabilistic functions of finite state Markov chains. Ann. Math. Stat. **37**(6), 1554–1563 (1966)
8. Huang, X., Acero, A., Hon, H.W.: Spoken Language Processing: A Guide to Theory, Algorithm, and System Development. Prentice Hall (2001)
9. Cieri, C., Miller, D., Walker, K.: The Fisher Corpus: A Resource for the Next Generations of Speech-to-Text. In: International Conference on Language Resources and Evaluation (2004)
10. http://cmusphinx.sourceforge.net
11. Aarts, E., Korst, J.: Simulated Annealing and Boltzmann Machines. John Wiley and Sons Inc. (1988)

A Nonlinear Acoustic Echo Canceller with Improved Tracking Capabilities

Danilo Comminiello, Michele Scarpiniti, Simone Scardapane,
Raffaele Parisi and Aurelio Uncini

Abstract This paper introduces the use of a variable step size for a functional link adaptive filter (FLAF). We consider a split FLAF architecture, in which linear and nonlinear filterings are performed in two separate paths, thus resulting well-suited for online filtering applications, like the nonlinear acoustic echo cancellation (NAEC). We focus our attention on the nonlinear path to improve the overall NAEC performance. To this end, we derive a variable step size for the filter on the nonlinear path that shows reliance not only on the nonlinear path, but on the whole split FLAF architecture. The introduction of the variable step size for the nonlinear filter aims at improving the modeling of nonlinear speech signals, thus yielding superior performance in NAEC problems. Experimental results prove the effectiveness of the proposed method with respect to the standard split FLAF involving a fixed step size.

Keywords Functional links · Nonlinear acoustic echo cancellation · Nonlinear adaptive filtering · Nonlinear speech modeling

The work of Danilo Comminiello was partly funded by bdSound.

D. Comminiello (✉) · M. Scarpiniti · S. Scardapane · R. Parisi · A. Uncini
Department of Information Engineering, Electronics and Telecommunications (DIET)
"Sapienza" University of Rome, Via Eudossiana 18, 00184 Rome, Italy
e-mail: danilo.comminiello@uniroma1.it

M. Scarpiniti
e-mail: michele.scarpiniti@uniroma1.it

S. Scardapane
e-mail: simone.scardapane@uniroma1.it

R. Parisi
e-mail: raffaele.parisi@uniroma1.it

A. Uncini
e-mail: aurel@ieee.org

© Springer International Publishing Switzerland 2016
A. Esposito et al. (eds.), *Recent Advances in Nonlinear Speech Processing*,
Smart Innovation, Systems and Technologies 48,
DOI 10.1007/978-3-319-28109-4_24

235

1   Introduction

Nonlinear acoustic echo cancellation (NAEC) systems are widely used to model nonlinearities rebounding on acoustic echo paths that affect speech signals in hands-free communication systems. Such nonlinearities are mainly caused by loudspeakers and lead to a quality degradation of a speech communication. NAEC systems reduce the effect of nonlinearities, thus improving echo cancellation performance.

In this paper, we focus on a recently proposed NAEC system, which is based on the use of *functional link adaptive filters* (FLAFs) [5]. These filters are characterized by a nonlinear expansion of the input followed by a linear filtering of the expanded signal. In particular, we take into account a split FLAF (SFLAF) architecture [5], which separates the adaptation of linear and nonlinear elements in two parallel paths, each one devoted to a specific task. This structure is particularly significant in NAEC problems [4–6], since the linear path can be exclusively used to estimate the acoustic impulse response, while the nonlinear path can be committed to model any nonlinearity.

Usually, processing a speech signal is made difficult by its nonstationary nature. Moreover, a nonlinearity, like that produced by a loudspeaker, emphasizes the non-stationarity of a signal, such that modeling a distorted speech signal becomes very difficult. In order to improve the modeling of nonlinearities, the tracking performance of the nonlinear filter should be optimized according to the level of nonlinearity that affects a speech signal at each instant [10]. To this end, we propose the use of a variable step size for the adaptive filter on the nonlinear path of the SFLAF. Variable step sizes have been largely used for adaptive filters in linear system identification problems, such as acoustic echo cancellation and adaptive beamforming [1–3, 8, 11–13, 15]. However, in this paper the variable step size is used to provide improved tracking performance in the presence of nonlinear speech.

The rest of the paper is organized as follows: the SFLAF architecture is described in Sect. 2. In Sect. 3, a variable step size is introduced for the adaptive filter on the nonlinear path, and, in Sect. 4, experimental results are shown. Finally, in Sect. 5 our conclusions are presented.

2   The Split Functional Link Adaptive Filter

The *split functional link adaptive filter* (SFLAF) model [5], depicted in Fig. 1, is a parallel architecture including a linear path and a nonlinear path. The former is simply composed of a linear adaptive filter, which completely aims at modeling the linear components of an unknown system; the nonlinear path is composed of a Hammerstein cascade model comprising a functional expansion block and a subsequent adaptive filter.

At nth time instant the SFLAF receives the input sample $x[n]$, which is stored in the linear input buffer $\mathbf{x}_{L,n} \in \mathbb{R}^M = \left[x[n] \; x[n-1] \ldots x[n-M+1] \right]^T$, where M

Fig. 1 The split functional link adaptive filter

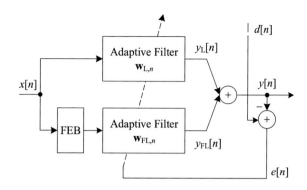

is the linear input buffer length, i.e., the length of linear adaptive filter $\mathbf{w}_{L,n} \in \mathbb{R}^M = \left[w_{L,0}\,[n]\ w_{L,1}\,[n] \ \ldots \ w_{L,M-1}\,[n] \right]^T$. The adaptive filtering yields the linear output $y_L\,[n] = \mathbf{x}_{L,n}^T \mathbf{w}_{L,n-1}$. On the other hand, the nonlinear path receives a subvector of $\mathbf{x}_{L,n}$ as nonlinear input buffer: $\mathbf{x}_{FL,n} \in \mathbb{R}^{M_i} = \left[x\,[n]\ x\,[n-1] \ \ldots \ x\,[n-M_i+1] \right]^T$, where $M_i \leq M$ is defined as the nonlinear input buffer length, which can be equal to the linear input buffer length or just a portion of it. The nonlinear path receives the nonlinear buffer $\mathbf{x}_{FL,n}$, which is processed by means of a *functional expansion block* (FEB). The FEB consists of a series of functions, which might be a subset of a complete set of orthonormal basis functions, satisfying universal approximation constraints. The term "functional links" refers to the functions contained in the chosen set $\Phi = \{\varphi_0\,(\cdot)\,, \varphi_1\,(\cdot)\,, \ldots, \varphi_{Q_f-1}\,(\cdot)\}$, where Q_f is the number of functional links. The FEB processes the input buffer by passing each element of the buffer $\mathbf{x}_{FL,n}$ as argument for the chosen functions, each one yielding a subvector $\bar{\mathbf{g}}_{i,n} \in \mathbb{R}^{Q_f}$:

$$\bar{\mathbf{g}}_{i,n} = \left[\varphi_0\,(x\,[n-i])\ \varphi_1\,(x\,[n-i]) \ \ldots \ \varphi_{Q_f-1}\,(x\,[n-i]) \right]. \tag{1}$$

The concatenation of such subvectors yields an *expanded buffer* $\mathbf{g}_n \in \mathbb{R}^{M_e}$:

$$\mathbf{g}_n = \left[\bar{\mathbf{g}}_{0,n}^T\ \bar{\mathbf{g}}_{1,n}^T \ \ldots \ \bar{\mathbf{g}}_{M_i-1,n}^T \right]^T \tag{2}$$

where $M_e = Q_f \cdot M_i \geq M_i$ represents the length of the expanded buffer. Note that $M_e = M_i$ only when $Q_f = 1$. The functional expansion chosen for this work is a nonlinear trigonometric series expansion such that:

$$\varphi_j\,(x\,[n-i]) = \begin{cases} \sin\,(p\pi x\,[n-i])\,, & j = 2p-2 \\ \cos\,(p\pi x\,[n-i])\,, & j = 2p-1 \end{cases} \tag{3}$$

where $p = 1, \ldots, P$ is the expansion index, being P the *expansion order*, and $j = 0, \ldots, Q_f - 1$ is the functional link index. Therefore, in the case of trigonometric expansion, the functional link set Φ is composed of $Q_f = 2P$ functional links. Convergence performance of a trigonometric FLAF is shown in [9]. Note that (3) actually refers to a *memoryless* expansion, since it does not involve cross-products of the nth input sample with previous samples. However, the same process holds also for functional expansion with memory. The choice of involving some memory may be decisive when the nonlinearity introduced by the system to be identified is characterized by a dynamic nature, i.e., depends also on the time instant. In our model, we consider the memory of a nonlinearity by taking into account the outer products of the ith input sample with the functional links of the previous K input samples, where K represents the *memory order* (see [5] for a detailed explanation).

The achieved expanded buffer $\mathbf{g}_n$ is then fed into a linear adaptive filter $\mathbf{w}_{\text{FL},n} \in \mathbb{R}^{M_e} = \left[w_{\text{FL},0}[n]\ w_{\text{FL},1}[n]\ \ldots\ w_{\text{FL},M_e-1}[n] \right]^T$, thus providing the nonlinear output $y_{\text{FL}}[n] = \mathbf{g}_n^T \mathbf{w}_{\text{FL},n-1}$. The SFLAF output results from the sum of the two path outputs:

$$y[n] = y_{\text{L}}[n] + y_{\text{FL}}[n] \tag{4}$$

and, thereby, the overall error signal[1] is:

$$e[n] = d[n] - y[n] = d[n] - \mathbf{x}_{\text{L},n}^T \mathbf{w}_{\text{L},n-1} - \mathbf{g}_n^T \mathbf{w}_{\text{FL},n-1}, \tag{5}$$

which is used for the adaptation of both adaptive filters. In (5), $d[n]$ represents the desired signal including any near-end additive noise $v[n]$ and a near-end speech contribution $s[n]$. We use a standard *normalized least-mean square* (NLMS) algorithm (see for example [14, 16]) to adapt the coefficients of both $\mathbf{w}_{\text{L},n}$ and $\mathbf{w}_{\text{FL},n}$:

$$\mathbf{w}_{\text{L},n} = \mathbf{w}_{\text{L},n-1} + \mu_{\text{L}} \frac{\mathbf{x}_{\text{L},n} e[n]}{\mathbf{x}_{\text{L},n}^T \mathbf{x}_{\text{L},n} + \delta_{\text{L}}} \tag{6}$$

$$\mathbf{w}_{\text{FL},n} = \mathbf{w}_{\text{FL},n} + \mu_{\text{FL}}[n] \frac{\mathbf{g}_n e[n]}{\mathbf{g}_n^T \mathbf{g}_n + \delta_{\text{FL}}} \tag{7}$$

where δ_{L} and δ_{FL} are regularization factors, and μ_{L} is a fixed step size for the filter on the linear path, and $\mu_{\text{FL}}[n]$ is the variable step size parameter, on which we focus in the next section in order to improve the nonlinear modeling performance.

3   A Variable Step Size for the Nonlinear FLAF

In this section, we derive the *variable step size* $\mu_{\text{FL}}[n]$ of (7), thus providing a reliable solution to nonlinear speech modeling. In order to yield an algorithm easy to control in practical implementations, similarly to what done in [12] for linear echo cancellation, the derivation is taken considering that no *a priori* information must be required about the nonlinearity to be modeled.

---

[1] It may also be denoted as *a priori* output estimation error [14] to be distinguished from the *a priori* estimation error, which is defined as $e_a[n] = e[n] - v[n]$, where $v[n]$ is additive noise.

We start from the consideration that the desired signal $d[n]$ is composed of a signal $\tilde{x}[n]$, which is generated by the far-end signal convolved with an acoustic impulse response and distorted by any nonlinear process. The desired signal also contains a near-end contribution $s[n]$ and any additive noise $v[n]$, therefore, it can be written as:

$$d[n] = \tilde{x}[n] + s[n] + v[n] \tag{8}$$

In order to derive the optimal variable step size parameter, we assume that $\tilde{x}[n]$, $s[n]$ and $v[n]$ are statistically uncorrelated and we take the squares and the expectations of both sides of (8), thus resulting:

$$E\left\{d^2[n]\right\} = E\left\{\tilde{x}^2[n]\right\} + E\left\{s^2[n]\right\} + E\left\{v^2[n]\right\} \tag{9}$$

According to the *least perturbation property* [14], at steady state, i.e., for $n \to \infty$, the weights of an adaptive filter no longer change during the adaptation. Therefore, it is reasonable to assume the following approximation:

$$E\left\{\tilde{x}^2[n]\right\} \approx E\left\{y^2[n]\right\} + E\left\{q^2[n]\right\}. \tag{10}$$

In (10) an irreducible noise term $q[n]$ has been introduced due to the nonlinear approximation. Therefore, Eq. (9) turns into the following one:

$$E\left\{d^2[n]\right\} - E\left\{y^2[n]\right\} = E\left\{s^2[n]\right\} + E\left\{v^2[n]\right\} + E\left\{q^2[n]\right\}. \tag{11}$$

The right member of (11) contains the near-end contribution and the irreducible noise, that may be approximated to the *a posteriori* output estimation error $\varepsilon[n]$ at steady state [3, 14]. Therefore, Eq. (11) can be written as:

$$E\left\{d^2[n]\right\} - E\left\{y^2[n]\right\} \approx E\left\{\varepsilon^2[n]\right\}. \tag{12}$$

However, in order to achieve the optimal $\mu_{\text{FL}}[n]$, we need to express $\varepsilon[n]$ in terms of the *a priori* error $e[n]$. A relation between the *a posteriori* and *a priori* error signals may be derived starting from the definition of $\varepsilon[n]$:

$$\varepsilon[n] = d[n] - \mathbf{x}_n^T \mathbf{w}_{\text{L},n} - \mathbf{g}_n^T \mathbf{w}_{\text{FL},n} \tag{13}$$

Replacing the update Eqs. (6) and (7) in (13), and taking into account the *a priori* error signal definition (5), it is possible to achieve the following relation:

$$\varepsilon[n] = (1 - \mu_{\text{L}} - \mu_{\text{FL}}[n]) e[n] \tag{14}$$

The step size μ_{L} in (14) may be a fixed value or even a variable parameter achieved by any variable step size technique. However, the goal of the paper is to investigate the effects of a variable step size for the nonlinear modeling and thus we consider μ_{L} as a fixed value. Therefore, we can replace Eq. (14) in (12), thus resulting:

$$\mathrm{E}\left\{d^2\left[n\right]\right\} - \mathrm{E}\left\{y^2\left[n\right]\right\} = \left|1 - \mu_L - \mu_{FL}\left[n\right]\right|^2 \mathrm{E}\left\{e^2\left[n\right]\right\} \tag{15}$$

from which we can derive an expression of the variable step size parameter $\mu_{FL}\left[n\right]$:

$$\mu_{FL}\left[n\right] = \left|1 - \mu_L - \sqrt{\frac{\mathrm{E}\left\{d^2\left[n\right]\right\} - \mathrm{E}\left\{y^2\left[n\right]\right\}}{\mathrm{E}\left\{e^2\left[n\right]\right\}}}\right|. \tag{16}$$

From a practical point of view, we evaluate the expectations in terms of power estimates, as done for example in [12], thus achieving:

$$\mu_{FL}\left[n\right] = \left|1 - \mu_L - \frac{\sqrt{\left|\widehat{\sigma}_d^2\left[n\right] - \widehat{\sigma}_y^2\left[n\right]\right|}}{\widehat{\sigma}_e^2\left[n\right] + \xi}\right|. \tag{17}$$

In (17), the general parameter $\widehat{\sigma}_\theta^2\left[n\right]$ represents the power estimate of the sequence $\theta\left[n\right]$, being $\theta = \{d, y, e\}$, and it can be computed as:

$$\widehat{\sigma}_\theta^2\left[n\right] = \beta\widehat{\sigma}_\theta^2\left[n - 1\right] + (1 - \beta)\,\theta^2\left[n\right] \tag{18}$$

where β is a forgetting factor, whose value can be chosen as $\beta = 0.99$. A small positive number $\xi = 10^{-4}$ is added in (17) to avoid divisions by zero. Another practical consideration is that, in presence of high dynamic nonlinearities, the power of the estimate of the output signal $\widehat{\sigma}_y^2\left[n\right]$ may be larger than the power of the desired signal $\widehat{\sigma}_d^2\left[n\right]$; this is the reason why the absolute value of the terms under the square root is considered.

4 Experimental Results

We assess the effectiveness of the proposed FLAF-based architecture in a nonlinear acoustic echo cancellation problem. Experiments take place in a simulated room environment with a reverberation time of $T_{60} \approx 100$ ms measured at 8 kHz sampling frequency. A far-end signal $x\left[n\right]$ is reproduced by a simulated loudspeaker and captured by a microphone. In order to have a complete view of the effects of the nonlinearity, we use both a colored noise and a speech signal as far-end input. The colored noise signal is generated by means of a first-order autoregressive model, whose transfer function is $\sqrt{1 - \theta^2}/\left(1 - \theta z^{-1}\right)$, with $\theta = 0.8$, fed with an independent and identically distributed (i.i.d.) Gaussian random process. In order to simulate a loudspeaker distortion, we apply a symmetrical soft-clipping nonlinearity to the far-end signal [6, 7]:

Fig. 2 Scheme of the NAEC
system

$$\bar{y}[n] = \begin{cases} \frac{2}{3\zeta}x[n] & , \ 0 \le |x[n]| \le \zeta \\ \text{sign}(x[n])\frac{3-(2-|x[n]|/\zeta)^2}{3} & , \ \zeta \le |x[n]| \le 2\zeta \\ \text{sign}(x[n]) & , \ 2\zeta \le |x[n]| \le 1 \end{cases} \qquad (19)$$

where $0 < \zeta \le 0.5$ is a nonlinearity threshold. As also described by Fig. 2, the signal
$\bar{y}[n]$ is then convolved with an acoustic impulse response related to the simulated
room environment, thus achieving the desired signal $d[n]$ acquired by a microphone.
The signal $d[n]$ contains also near-end background noise $v[n]$, in the form of additive
Gaussian noise, providing 20 dB of *signal-to-noise ratio* (SNR). The length of the
acoustic impulse response is $M = 300$.

Performance is evaluated in terms of the *echo return loss enhancement* (ERLE),
expressed in dB as: ERLE $[n] = 10\log_{10}\left(\text{E}\left\{d^2[n]\right\}/\text{E}\left\{e^2[n]\right\}\right)$. We use the following parameter setting: input buffer length $M_i = M$, fixed step-size parameter
$\mu_L = 0.2$, regularization parameter $\delta_{FL} = 10^{-2}$ for both the filters of the SFLAF,
expansion order $P = 10$, memoryless functional links (i.e., $K = 0$), and distortion
threshold $\zeta = 0.15$. We compare the results of the proposed VSS-SFLAF with a
SFLAF having the same parameter setting of the VSS-SFLAF, but a fixed step-size
value $\mu_{FL} = 0.2$.

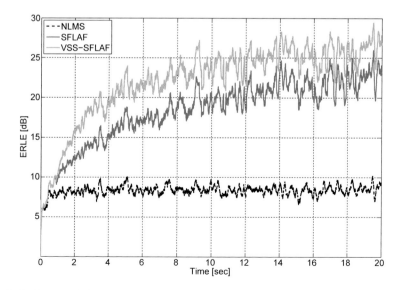

Fig. 3 Performance behavior in terms of ERLE in case of colored noise input

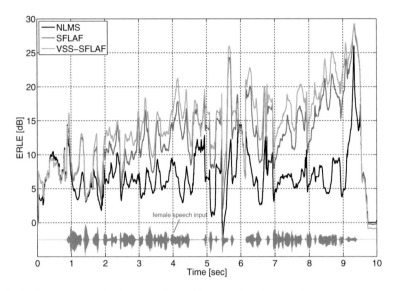

Fig. 4 Performance behavior in terms of ERLE in case of female speech input

In case of colored noise, the VSS-SFLAF achieves a good improvement in terms of tracking performance over the SFLAF, as it is possible to see in Fig. 3, while tending to a similar behavior at steady state. Results becomes more significant when using an input signal with a high nonstationarity, i.e., a speech signal, since performance improvements are more difficult to be obtained in this case. As depicted in Fig. 4, a gain over the SFLAF can be achieved, not only in proximity of the peaks of the speech signal, but for the whole length of the signal.

5 Conclusions

In this paper, a functional link-based nonlinear acoustic echo canceller is proposed involving a variable step size on the nonlinear path of the architecture. The proposed VSS-SFLAF takes advantage from the use of the variable step size, thus improving the tracking performance of nonlinear speech signals. Future research will include the use of a joined variable step size that governs the convergence performance of both filters on the linear and nonlinear paths.

References

1. Aboulnasr, T., Mayyas, K.: A robust variable step-size LMS-type algorithm: analysis and simulations. IEEE Trans. Signal Process. **45**(3), 631–639 (1997)
2. Albu, F., Coltuc, D., Comminiello, D., Scarpiniti, M.: The variable step size regularized block exact affine projection algorithm. In: Proceedings of the IEEE International Symposium on Electronics and Telecommunications (ISETC), pp. 283–286. Timisoara, Romania, Nov 2012
3. Benesty, J., Rey, H., Vega, L.R., Tressens, S.: A nonparametric VSS NLMS algorithm. IEEE Signal Process. Lett. **13**(10), 581–584 (2006)
4. Comminiello, D., Azpicueta-Ruiz, L.A., Scarpiniti, M., Uncini, A., Arenas-García, J.: Functional link based architectures for nonlinear acoustic echo cancellation. In: Proceedings of the IEEE Joint Workshop on Hands-free Speech Communication and Microphone Arrays (HSCMA), pp. 180–184. Edinburgh, UK, May 2011
5. Comminiello, D., Scarpiniti, M., Azpicueta-Ruiz, L.A., Arenas-García, J., Uncini, A.: Functional link adaptive filters for nonlinear acoustic echo cancellation. IEEE Trans. Audio Speech Lang. Process. **21**(7), 1502–1512 (2013)
6. Comminiello, D., Scarpiniti, M., Azpicueta-Ruiz, L.A., Arenas-García, J., Uncini, A.: Nonlinear acoustic echo cancellation based on sparse functional link representations. IEEE/ACM Trans. Audio Speech Lang. Process. **7**(22), 1172–1183 (2014)
7. Comminiello, D., Scardapane, S., Scarpiniti, M., Parisi, R., Uncini, A.: Functional Link Expansions for Nonlinear Modeling of Audio and Speech Signals. In: Proceedings of the IEEE International Joint Conference on Neural Networks (IJCNN), pp. 1–8. Killarney, Ireland, Jul 2015
8. Comminiello, D., Scarpiniti, M., Parisi, R., Uncini, A.: A novel affine projection algorithm for superdirective microphone array beamforming. In: Proceedings of the IEEE International Symposium on Circuits and Systems (ISCAS), pp. 2127–2130. Paris, France, May 2010
9. Comminiello, D., Scarpiniti, M., Parisi, R., Uncini, A.: Convergence properties of nonlinear functional link adaptive filters. IET Electron. Lett. **49**(14), 873–875 (2013)
10. Comminiello, D., Scarpiniti, M., Scardapane, S., Parisi, R., Uncini, A.: Improving nonlinear modeling capabilities of functional link adaptive filters. Neural Networks **69**, 51–59 (2015)
11. Huang, H.C., Lee, J.: A new variable step-size nlms algorithm and its performance analysis. IEEE Trans. Signal Process. **60**(4), 2055–2060 (2012)
12. Paleologu, C., Benesty, J., Ciochinǎ, S.: A variable step-size affine projection algorithm designed for acoustic echo cancellation. IEEE Trans. Audio Speech Lang. Process. **16**(8), 1466–1478 (2008)
13. Vega, Rey: L., Rey, H., Benesty, J., Tressens, S.: A new robust variable step-size nlms algorithm. IEEE Trans. Signal Process. **56**(5), 1878–1893 (2008)
14. Sayed, A.H.: Adaptive Filters. John Wiley & Sons, Hoboken, NJ (2008)
15. Shin, H.C., Sayed, A.H., Song, W.J.: Variable step-size NLMS and adffine projection algorithms. IEEE Signal Process. Lett. **11**(2), 132–135 (2004)
16. Uncini, A.: Fundamentals of Adaptive Signal Processing. Springer International Publishing AG, Cham, Switzerland, Signal and Communication Technology (2015)

Part VI
The Social Life of Speech Features

When the Game Gets Difficult, then it is Time for Mimicry

Vijay Solanki, Alessandro Vinciarelli, Jane Stuart-Smith and Rachel Smith

Abstract The computing community shows significant interest for the detection of mimicry, one of the names designating the tendency of interacting people to converge towards common behavioural patterns. This work shows experiments where speaker verification techniques, originally designed to detect fraudulent attempts to imitate others, are used to automatically detect the phenomenon. Furthermore, the experiments show that mimicry tends to be more frequent when people deal with harder collaborative tasks, thus suggesting that one of the functions of the phenomenon is to make communication easier or more effective in case of difficulties.

Keywords Mimicry · Social Signal Processing · Mixtures of Gaussians

1 Introduction

Automatic analysis of social and psychological phenomena taking place in conversations has attracted significant attention in recent years, especially after the development of domains like Social Signal Processing [20] and Computational Paralinguistics [18]. In particular, the computing community has made significant efforts aimed at automating the analysis of a large number of social phenomena like, e.g., dominance [14], personality traits [21], conflict [7], emotions [22], and roles [17] (see [21] for an extensive survey).

V. Solanki · A. Vinciarelli (✉) · J. Stuart-Smith · R. Smith
University of Glasgow, Glasgow G12 8QQ, UK
e-mail: Alessandro.Vinciarelli@glasgow.ac.uk

V. Solanki
e-mail: Vijay.Solanki@glasgow.ac.uk

J. Stuart-Smith
e-mail: Jane.Stuart-Smith@glasgow.ac.uk

R. Smith
e-mail: Rachel.Smith@glasgow.ac.uk

© Springer International Publishing Switzerland 2016
A. Esposito et al. (eds.), *Recent Advances in Nonlinear Speech Processing*,
Smart Innovation, Systems and Technologies 48,
DOI 10.1007/978-3-319-28109-4_25

Mimicry is one of the phenomena that has attracted most attention (see [11] for an extensive survey). The literature adopts different names—e.g. accommodation [12], interpersonal adaptation [9], synchrony (in particular when the convergence concerns temporal behavioural patterns) [10], etc.—but they all refer to the unconscious tendency of people to imitate others they interact with. For this reason, this article proposes the use of speaker verification approaches—originally conceived to detect fraudulent attempts to reproduce voice and speech of others [6]—to detect mimicry.

The key-idea of the approach applies to dyadic conversations involving two speakers A and B. If $p(X|\Theta_A)$ and $p(X|\Theta_B)$ are statistical models of the acoustic evidence produced by A and B (X is a sequence of observation vectors extracted from an utterance, Θ_A and Θ_B are the parameter sets of the two models), then the likelihoods $p(X_B|\Theta_A)$ and $p(X_A|\Theta_B)$ should improve when there is mimicry (where X_Y is a sequence of observation vectors extracted from speech samples uttered by speaker Y).

Preliminary experiments show that the improvement of the likelihoods above can be actually observed and, in a large number of cases, it is statistically significant. Furthermore, the experiments show that statistically significant improvements tend to be more frequent when the subjects speak longer. In the interaction scenario adopted for the experiments, this means that the subjects find it more difficult to address a task they have been assigned. Thus, the results seem to suggest not only that the approach detects and measures mimicry, but also that the function of mimicry (or at least one of its functions) is to make interaction easier during difficult collaborative efforts.

The rest of the paper is organized as follows: Sect. 2 introduces the mimicry problem from a linguistic and psychological point of view, Sect. 3 introduces data and scenario, Sect. 4 describes the approach proposed in this work, Sect. 5 reports on experiments and results, and Sect. 6 draws some conclusions.

2 Previous Work

Communicating with another person is a shared experience which necessitates the co-operation of both parties involved. This interactive, co-operative process requires some give and take from the speakers in order to ensure successful communication. Speakers have been observed to change their speech patterns in relation to the qualities of their conversational partner. Within the field of linguistics, this phenomenon was termed speech accommodation [12]. It has been found that, when exposed to the speech of another person, speakers shift their speech patterning either towards or away from that which they are presented with. In more recent years, it has been discovered that this phenomenon can be found at the fine grained level of phonetic changes [3, 15, 16, 19]. Generally speaking, speakers have been shown to modulate the realisation of both segmental (eg. VOT, $f1/f2$ in vowels) and supra-segmental (eg. pitch, speech rate) features of speech in response to being exposed to an excerpt of speech. The vast majority of studies of this kind tend to base their findings on small excerpts of carefully controlled stimuli, allowing for a great deal of experimental control at a fine grained level. This approach has been of great benefit as the findings demonstrate a robust accommodative effect even when subjects are given minimal information from the speech signal to accommodate to.

Additionally, these studies have mostly focused on demonstrating that accommodation can be found in single elements of a person's phonetic repertoire (i.e. only VOT or only $f1/f2$ in vowels). Again, this has provided an excellent grounding for the investigation of this phenomenon as it has been demonstrated to manifest in a wide array of phonetic elements. However, recently there has been a move to begin to assess accommodation in a more holistic manner. For instance, [4] assessed the degree to which listeners accommodated to the fundamental frequency (pitch / $f0$) of a male speaker and indeed found evidence for accommodation as measured by degree of fundamental frequency similarity. They also investigated the degree to which people can perceive accommodation to fundamental frequency. Whilst accommodation was detected, the results did not correlate with those collected from the first experiment. They conclude by highlighting that phonetic accommodation as assessed from a perceptual standpoint constitutes a number of phonetic features and is likely to be perceived holistically.

Further to this, there is also a move to begin to capture accommodation in dyadic interactions rather than in response to pre-recorded stimuli [1]. Attempting to combine both a holistic measure of phonetic accommodation with dyadic data using traditional analyses as applied by phoneticians would be very costly and labour intensive. All the phonetic elements of a conversation would need to be transcribed and analysed manually. The results would then need to be collected and catalogued into a searchable corpora with information not only about the phonetic variant in question but also about the surrounding phonetic environment. Whilst there are some tools designed to help in transcription[1] and analysis [13], they are still not wholly trusted by the community and can only partially aid in the work. An alternative approach would be to develop a holistic measure which captures changes in phonetic variance of both speakers over time.

3   Data and Scenario

The experiments contained in this work were performed on a corpus of 6 conversations (12 unacquainted participants in total) elicited using the Diapix UK scenario [5]. It consists of twelve images, each with a counterpart that is the same apart from twelve slight differences. Participants are tasked with finding the differences between the images, using verbal communication only. Here, two unacquainted participants were required to find all of the differences within 15 min. In order to limit the effect of

---

[1] FAVE (Forced Alignment and Vowel Extraction) Program Suite, I. Rosenfelder, J. Fruehwald, K. Evanini and Y. Jiahong.

non-verbal communication as much as possible, the participants sat in opposite corners of a sound attenuated booth, with a divider between them. They could not see each other but could still hear one another.

Overall, the 6 pairs spoke for 9 h and 37 min. Each pair went through all the 12 images of the Diapix Task, this allows the corpus to be split into $12 \times 6 = 72$ intervals, each corresponding to one of the pictures (the average duration is 8 min and 1 s). The task images were randomised before presentation to the different pairs. In this way, effects due to the position of a picture (e.g., tiredness effects for the pictures addressed at the end) should be limited. The participants were all females born and raised in the Glasgow conurbation. The main reason behind this choice is that gender and accent have been shown to play a role in mimicry [2, 8]. The age range is 19 to 65 with an average of 30.9.

4 The Approach

The approach proposed in this work includes three main steps. The first is the segmentation of the speech stream into words (this task is currently implemented manually), the second is the conversion of each word into a sequence of feature vectors (in the case of this work, the first 12 Mel Frequency Cepstral Coefficients), the third is the actual detection of mimicry. Given that the first step is performed manually, the rest of this section focuses only on the latter two.

4.1 Feature Extraction

Every word uttered during the conversations of the corpus is converted into a sequence of 12-dimensional MFCC vectors. The features are extracted at regular time steps of 10 *ms* from 30 *ms* long analysis windows. The reasons for using such a feature extraction process are mainly two. On the one hand, MFCCs have been found to be effective in speaker verification tasks [6]. On the other hand, MFCC account for speaker independent aspects of speech towards which the subjects can actually converge.

4.2 Mimicry Detection

Figure 1 shows the main elements of the approach. The conversations are first segmented into intervals that correspond to one of the Diapix pictures (see Sect. 3). For a given interval, all the words are converted into sequences of 12-dimensional MFCC vectors $X_i^{(A)}$ and $X_k^{(B)}$, where A and B are the two speakers involved in the same conversation. The words have been segmented manually and it is possible to know

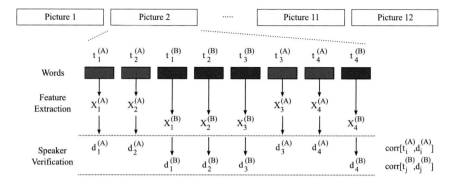

Fig. 1 The figure shows the experimental approach proposed in this work

their start times $t_i^{(A)}$ and $t_k^{(B)}$ (the words uttered by the two speakers are analyzed separately).

For every word i of speaker A, it is possible to estimate the following likelihood ratio:

$$d_i^{(A)} = \frac{p(X_i^{(A)}|\Theta_B)}{p(X_i^{(A)}|\Theta_A)}, \qquad (1)$$

where $X_i^{(A)}$ is the sequence of vectors extracted from the word, and Θ_A and Θ_B are the parameter sets of the models trained over A and B. By switching A and B in the equation above, it is possible to define $d_k^{(B)}$ for the words uttered by B.

In the experiments of this work, $p(X|\Theta_A)$ and $p(X|\Theta_B)$ are estimated using Mixtures of Gaussians:

$$p(X|\Theta) = \prod_{l=1}^{|X|} \sum_{n=1}^{G} \pi_n \mathcal{N}(x_l|\mu_n, \Sigma_n) \qquad (2)$$

where $|X|$ is the number of vectors in $\mathbf{x}$, G is the number of Gaussians in the mixture, π_l is the mixing coefficient of Gaussian l in the mixture, and Σ_l is the covariance matrix in the same Gaussian (the speaker index has been dropped for clarity).

5 Experiments and Results

Since every conversation can be segmented into 12 intervals (see Fig. 1), the Mixtures of Gaussians have been trained over the first interval of each conversation and used for testing over the rest of the data. In this way, the test set corresponds to $11 \times 6 = 66$ segments that can be further multiplied by 2 because words uttered by different

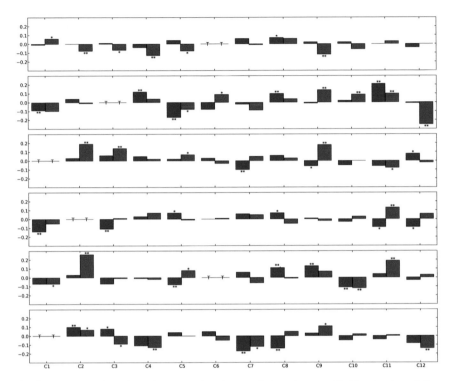

Fig. 2 The charts show, for ach conversation the correlations between likelihood ratio and time observed for each Diapix picture. The two bars for every picture correspond to the two speakers. *Asterisk* and double *asterisk* correspond to significance level 0.05 and 0.01, respectively

speakers are considered separately. As a result, the test set includes $66 \times 2 = 132$ samples.

For every sample that includes words uttered by A, it is possible to estimate the correlation between $d_i^{(A)}$ and $t_i^{(A)}$. If the correlation is statistically significant and positive, it means that A tends to get closer to B and, hence, A tends to mimick B. Viceversa, if the correlation is negative and statistically significant, it means that A tends to diverge from B. When the correlation is not statistically significant, there is no effect. The same considerations apply to B and the correlation between $d_k^{(B)}$ and $t_k^{(B)}$.

Figure 2 shows, for each pair of subjects, the correlations measured over the 11 time intervals used for test. The two bars for a given time interval correspond to the two speakers. In this way, it is possible to test whether each of the subjects converges individually towards her interlocutor. Out of 132 correlations, 51 are statistically significant with confidence level $\alpha = 0.05$ or less. The probability of this happening by chance, estimated with a two-tailed binomial test, is 10^{-12}.

Observing a large number of statistically significant correlations, well beyond what is expected by chance, suggests that the approach actually distinguishes between

Table 1 The table reports the average time (± the standard error) required for completing a task in Positive (the largest significant correlation is positive), Negative (the largest significant correlation is negative) and Null (there is no significant correlation) condition

Condition	Positive	Negative	Null
Avg. Length (s)	585 ± 51	427 ± 43	430 ± 33

The Positive condition is associated to tasks that require longer time to be addressed

intervals over which mimicry takes place and intervals over which it does not. However, further confirmation can come from the relationship between the outcome of the approach and an independent observed variable. Table 1 shows the length of the time intervals where statistically significant positive correlation is observed for at least one of the two speakers. In other words, the figure shows the duration of those intervals where, according to the approach proposed in this work, at least one of the two speakers converges towards the other. The chart clearly shows that mimicry tends to take place more frequently when spotting the differences in a picture takes more time.

The data of Table 1 can be interpreted in two ways: the first is that the more two people talk together, the more they tend to imitate one another. The second is that mimicry tends to take place more frequently when the subjects find it difficult to spot the differences. If the first explanation was true, mimicry should be more frequent when the subjects deal with the last pictures (at that point, the speakers have been interacting for long time). However, this is not what can be observed in Fig. 2 and, therefore, the second explanation appears to be more likely.

6 Conclusions

The experiments of this work show that methodologies inspired by speaker verification can help to detect mimicry, i.e. the unconscious tendency to imitate others during social interactions. Furthermore, the experiments show that the proposed approach tends to detect mimicry more frequently when the subjects speak longer. In the particular scenario adopted for the experiments, this means that the subjects find it difficult to address the task they are assigned. Thus, the experiments suggest that the function of mimicry is, at least in the case of the corpus used in this work, to help subjects when they are in difficulty.

Still, the experiments of this work are preliminary and the speaker verification methodologies adopted are basic. Therefore, future work includes two main directions. The first is to adopt different feature extraction techniques to account for different aspects of speech. The second is to move beyond Mixtures of Gaussians to use Hidden Markov Models that take into account temporal aspects and, possibly, the word being uttered.

254

V. Solanki et al.

References

1. Aguilar, L., Downey, G., Krauss, R., Pardo, J., Lane, S., Bolger, N.: A dyadic perspective on speech accommodation and social connection: Both partners' rejection sensitivity matters. J. Personal. (2015)
2. Babel, M.: Dialect divergence and convergence in New Zealand english. Lang. Soc. **39**(04), 437–456 (2010)
3. Babel, M.: Evidence for phonetic and social selectivity in spontaneous phonetic imitation. J. Phon. **40**(1), 177–189 (2012)
4. Babel, M., Bulatov, D.: The role of fundamental frequency in phonetic accommodation. Lang. Speech **55**(2), 231–248 (2012)
5. Baker, R., Hazan, V.: DiapixUK: task materials for the elicitation of multiple spontaneous speech dialogs. Behav. Res. Methods **43**(3), 761–770 (2011)
6. Bimbot, F., Bonastre, J.F., Fredouille, C., Gravier, G., Magrin-Chagnolleau, I., Meignier, S., Merlin, T., Ortega-García, J., Petrovska-Delacrétaz, D., Reynolds, D.: A tutorial on text-independent speaker verification. EURASIP J. Appl. Signal Process. **2004**, 430–451 (2004)
7. Bousmalis, K., Mehu, M., Pantic, M.: Towards the automatic detection of spontaneous agreement and disagreement based on nonverbal behaviour: A survey of related cues, databases, and tools. Image Vis. Comput. **31**(2), 203–221 (2013)
8. Bulatov, D.: The effect of fundamental frequency on phonetic convergence. Phonology lab annual report, University of California at Berkeley (2009)
9. Burgoon, J.K., Stern, L.A., Dillman, L.: Interpersonal Adaptation: Dyadic interaction patterns. Cambridge University Press, Cambridge (1995)
10. Chetouani, M.: Role of inter-personal synchrony in extracting social signatures: Some case studies. In: Proceedings of the Workshop on Roadmapping the Future of Multimodal Interaction Research Including Business Opportunities and Challenges, pp. 9–12 (2014)
11. Delaherche, E., Chetouani, M., Mahdhaoui, A., Saint-Georges, C., Viaux, S., Cohen, D.: Interpersonal synchrony: a survey of evaluation methods across disciplines. IEEE Trans. Affect. Comput. **3**(3), 349–365 (2012)
12. Giles, H., Coupland, J., Coupland, N.: Contexts of Accommodation: Developments in Applied Sociolinguistics. Cambridge University Press, Cambridge (1991)
13. Henry, K., Sonderegger, M., Keshet, J.: Automatic measurement of positive and negative voice onset time. In: Proceedings of Interspeech (2012)
14. Jayagopi, D., Hung, H., Yeo, C., Gatica-Perez, D.: Modeling dominance in group conversations using nonverbal activity cues. IEEE Trans. Audio Speech Lang. Process. **17**(3), 501–513 (2009)
15. Pardo, J.S.: Reflections on phonetic convergence: speech perception does not mirror speech production. Lang. Linguist. Compass **6**(12), 753–767 (2012)
16. Purnell, T.C.: Convergence and contact in milwaukee: evidence from select african american and white vowel space features. J. Lang. Soc. Psychol. **28**(4), 408–427 (2009)
17. Salamin, H., Vinciarelli, A.: Automatic role recognition in multiparty conversations: an approach based on turn organization, prosody, and conditional random fields. IEEE Trans. Multimed. **14**(2), 338–345 (2012)
18. Schuller, B., Batliner, A.: Computational Paralinguistics: Emotion, Affect and Personality in Speech and Language Processing. Wiley (2013)
19. Tobin, S.: Phonetic accommodation in spanish-english and korean-english bilinguals. In: Proceedings of Meetings on Acoustics, vol. 19, p. 060087 (2013)
20. Vinciarelli, A., Pantic, M., Bourlard, H.: Social Signal Processing: survey of an emerging domain. Image Vis. Comput. J. **27**(12), 1743–1759 (2009)
21. Vinciarelli, A., Pantic, M., Heylen, D., Pelachaud, C., Poggi, I., D'Errico, F., Schroeder, M.: Bridging the gap between social animal and unsocial machine: a survey of Social Signal Processing. IEEE Trans. Affect. Comput. **3**(1), 69–87 (2012)
22. Zeng, Z., Pantic, M., Roisman, G., Huang, T.: A survey of affect recognition methods: audio, visual, and spontaneous expressions. IEEE Trans. Pattern Anal. Mach. Intell. **31**(1), 39–58 (2009)

Predicting Cognitive Load Levels
from Speech Data

Jing Su and Saturnino Luz

Abstract An analysis of acoustic features for a ternary cognitive load classification task and an application of a classification boosting method to the same task are presented. The analysis is based on a data set that encompasses a rich array of acoustic features as well as electroglottographic (EGG) data. Supervised and unsupervised methods for identifying constitutive features of the data set are investigated with the ultimate goal of improving prediction. Our experiments show that the different tasks used to elicit the speech for this challenge affect the acoustic features differently in terms of their predictive power and that different feature selection methods might be necessary across these sub-tasks. The sizes of the training sets are also an important factor, as evidenced by the fact that the use of boosting combined with feature selection was enough to bring the unweighted recall scores for the Stroop tasks well above a strong support vector machine baseline.

Keywords Paralinguistic information · Cognitive load modelling · Feature selection · Classification

1 Introduction

Non-verbal and paralinguistic characteristics of speech have received increasing attention from researchers. It is now commonly accepted that non-verbal sounds form an important part of human communication [3], and that non-verbal features may help identify important structural aspects of speech interaction [8] in both natural and laboratory settings [2, 9, 10]. A more recent trend in the use of paralinguistic

J. Su
Centre for Applied Data Analytics Research, University College Dublin,
Dublin, Ireland
e-mail: jing.su@ucd.ie

S. Luz (✉)
Usher Institute of Population Health Sciences & Informatics, The University of Edinburgh,
Edinburgh, UK
e-mail: S.Luz@ed.ac.uk; luzs@acm.org

© Springer International Publishing Switzerland 2016
A. Esposito et al. (eds.), *Recent Advances in Nonlinear Speech Processing*,
Smart Innovation, Systems and Technologies 48,
DOI 10.1007/978-3-319-28109-4_26

features is their analysis for predicting levels of cognitive workload. Determination of workload levels is relevant in fields such as ergonomics, where it could help improve human computer interaction [5]. While most research in this field has been based on neurophysiological measuring, which involves specialised and intrusive equipment, the use of voice features for assessment of cognitive load levels is seen as promising enough to motivate a COMputational PARalinguistic ChallengE, ComParE [11].

This paper comprises a study of supervised and unsupervised machine learning methods applied to the prediction of cognitive load levels on a dataset distributed as part of the ComParE' 14 dataset. As this dataset contains a large number of speech and electroglottographic features, we investigated unsupervised and supervised dimensionality reduction methods in order to eliminate contingent features of the data. We then trained ensembles of classifiers (using the boosting technique) in order to distinguish among the different (discretised) levels of cognitive load.

Experiments showed that the cognitive load prediction task is better handled with supervised feature selection and different classification schemes. Contrary to our expectations, principal component analysis (PCA) and Discrete Cosine Transform (DCT) feature extraction methods proved quite ineffective. However, with supervised feature selection a boosting global model achieved unweighted average recall (UAR) scores 20.5 and 18 % higher than a published baseline based on a tuned support vector machine (SVM) classifier [11], in a Stroop time pressure and dual task, respectively. Similar per-task models were not quite as successful, but still yielded an improvement of 12 % in the Stroop dual task data.

2 The Dataset

The Cognitive Load with Speech and EGG (CLSE) dataset [11, 14] was designed to support the investigation of acoustic features and evaluation of algorithms for the determination of a speaker's cognitive load and working memory during speech. The CLSE database comprises recordings of 20 male and 6 female native Australian English speakers. These recordings encompass four types of experimental tasks, namely: *reading span Sentence*, *reading span Letter*, *Stroop time pressure* and *Stroop dual task*. These tasks define four partitions of the CLSE dataset. In each case, the data instances are classified objectively into three distinct cognitive load levels: low (L1), medium (L2) and high (L3) levels.

The "span" tasks are used to measure the working memory capacity of a subject [14], in which participants are required to remember concepts or objects in the presence of distractors [4, 11]. The reading span task is based on the protocol described by Unsworth et al. [13, 14]. It required the participants to read a series of (between two to five) possibly illogical short sentences, indicate whether the sentence read was true or false, and then remember a single letter presented briefly between sentences. This setup allowed the gatherer of the dataset to label memory load levels objectively as: L1, for data from the first sentence, L2, for data from the second sentence, and L3, for data from the third, fourth, and fifth sentences (for which no further distinctions were made).

Table 1 Summary of instance quantities in each type of task

	Training	Validation	Test
Reading span letter	815	499	576
Reading span sentence	825	525	600
Stroop time pressure	99	63	72
Stroop dual task	99	63	72
Total	1838	1150	1320

The Stroop tasks (*Stroop time pressure* and *Stroop dual task*), named after JR Stroop's seminal experiments [12], aim to induce increased cognitive load through presentation of conflicting stimuli to the participant. In this case, the stimuli are word and colour. The participant is asked to name the font colour of words corresponding to different colour names. Data instances produced in conditions where both the colour and the word that named the colour were the same were labelled as L1 (low cognitive load). Where the font colours and the colour names differed, data were labelled L2 or L3 (medium or high level of cognitive load). The high level was defined in terms of the time pressure on the subject (i.e. the colour had to be named in a short period of time, namely 0.8 s) or in terms of task complexity (i.e. participants were required to perform a tone-counting task in addition to naming the font colour). These distinctions characterise the Stroop time pressure and Stroop dual task subsets of the CLSE dataset.

Table 1 shows the standard "splits" of the CLSE dataset. The validation and the test set contain roughly same number of instances, while the training set contains about 50% more data. Among the four types of tasks employed in data collection, the two *span* tasks occupy the majority of the dataset while the two *Stroop* tasks comprise only about 10% of each dataset. Considering that the dataset has 6,374 attributes in total, one can readily see that the *Stroop* sets are affected more severely by high dimensionality.

A fair portion of features in the training set have very low variance. This includes, for instance, all quadratic regression coefficients of level 1, and a number of other prosodic features. Some low level descriptors of spectral features also suffer from this problem. These features are nearly constant and bring little discriminatory power to the classification model. We therefore removed all features with standard deviation less than 0.01. In Total 252 features (3.95% of all features) were removed from the training set, as a preprocessing step for all modelling experiments in this paper.

3 Predicting Cognitive Load Labels

A training set containing 1,838 instances described by 6,374 features challenges most classifiers since the data points are sparse with respect to dimensionality. The sparsity is more severe for models trained on subsets that contain only instances of a particular task (per task models). We therefore started by assessing the potential of three dimensionality reduction methods in rendering the dataset more tractable by learning algorithms.

3.1 PCA Experiments

PCA seeks to reduce dimensionality while preserving most of data variation. Applying PCA to a dataset transformed so that all features are scaled and centered, we found that the first eight principal components explain over 95 % of cumulative variance. We took 20 PCs and reencoded training and validation sets into this new space. The cleaned features are projected onto the 20 PCs, and used for training (the transformed training set has 1,838 instances with 20 features). When testing with the validation set, features need to be projected to the 20 PCs before the prediction step.

Here a global model is trained and used to predict on each instance in the validation set. UAR scores were collected for each task. Contrary to our expectation, both a naive Bayes classifier and the AdaBoost classifier failed to produce satisfactory results. We found that the UAR scores were far below baseline with the SVM global model of [11]. We speculate that the reason of this low performance on the PCA-reduced sets is the lack of an effective method for normalising the data per speaker on the training and test set. In the absence of such normalisation, PCA may be dominated by a few predominant features which can easily lead this method to overfit.

3.2 DCT Experiments

Discrete Cosine Transform (DCT) expresses a finite sequence of data points as a weighted sum of cosine functions with different frequencies. DCT is similar to Discrete Fourier Transform (DFT) but only has real spectrum. DCT is widely used in image audio and video compression because it has strong energy compaction property [1] by which most signal information concentrates in a few low-frequency components.

In this study we apply DCT and inverse DCT transformation to the feature set before classification with a global model. In order to test the scale of effective components, a series of trials are made with 2 to 50 % low-frequency components. Together with DCT transformation, naive Bayes classifier, boosting classifiers with decision stump base learner and decision tree base learner as well as SVM classifiers with RBF kernel and polynomial kernel are tested with the validation set. However, none of the tests have higher UAR score than the baseline in each task.

Table 2 The effect of feature selection with AdaBoost classifier on validation set

	FS = No (%)	FS = Yes (%)	Baseline
Reading span sentence	48.50	55.39	61.3
Stroop time pressure	57.14	65.08	54.0
Stroop dual task	49.21	52.38	44.4

UAR scores are from the global model, and AdaBoost is trained 30 iterations with decision tree base classifier. FS indicates feature selection with the CfsSubsetEval filter

3.3 Feature Selection and Global Model

Faced with the failure of an unsupervised method of dimensionality reduction, we attempted a supervised approach. The CfsSubsetEval feature filter provided by the Weka package [6] was employed. It selects attributes by individual correlation with the class variable and inter-correlation with other attributes. Subsets of features that are highly correlated with the class while having low intercorrelation are preferred [7]. We compare global model prediction UAR scores with and without CfsSubsetEval pre-filtering in Table 2.

On classifier selection, we chose Boosting with a decision tree base learner rather than decision stump. Table 2 shows the efficacy of feature selection combined with an AdaBoost.M1 with Decision Tree base learner. Without feature selection, AdaBoostM1 beats the SVM baseline slightly in the Stroop tasks, but is 13 % lower than baseline in the reading task. This observation shows the power of ensemble classification in this dataset when there is a proper base learner. When feature selection is in use, the global model achieves higher accuracy for each task. In Stroop time pressure task, the best UAR is 65.08 %, an improvement of 11 points over the baseline. In the Stroop dual task, the best UAR is 52.38 %, an 8-point improvement over the baseline. However, reading span is still 6 % lower than baseline. In the next section we investigate per task models, where classifiers are trained on relatively more uniform training sets.

3.4 Per Task Model

In the above section, we predicted objective load level with a global model which trains a single model on all available instances and predicts on a validation set of each task. In this section we apply an alternative approach, training one model with data from one task and predicting on a validation set of the corresponding task. This is called a per task model [11].

The split training sets are filtered in the same way as for the above described experiments. Features with standard deviation less than 0.01 are pre-filtered. The CfsSubsetEval filter selects 93, 74 and 51 features by sequence for each task. Then AdaBoost.M1 is employed as a classifier for the corresponding per task models.

Table 3 The effect of feature selection with AdaBoost

	Ada+DT (%)	Ada+DS (%)	Baseline (%)
Reading span sentence	54.98	48.86	61.2
Stroop time pressure	68.25	73.02	74.6
Stroop dual task	66.67	71.43	63.5

UAR scores are from Per Task model, and AdaBoost is trained 20 iterations with each base learner

The number of training iterations is set to 20 for each base learner. Since the Decision Tree (DT) base learner works well for the Global model, it is used again. Moreover, we also use a Decision Stump (DS) base learner for comparison.

The results are shown in Table 3. Decision Stump, as the simplest tree structure, outperforms Decision Tree in AdaBoost for both Stroop tasks. This observation comes from per task model prediction on the validation set and seems quite surprising. In order to test its validity, we further analyse the Stroop Dual Task model prediction within the training set. Figure 1a shows the performance of both DS and DT base learners under different numbers of AdaBoost iterations. It is clear that AdaBoost with the DT base learner reaches 100 % UAR in the training set regardless of the number of training steps (10 to 100 iterations). At the same time, its prediction accuracy on the validation set oscillates between 61.90 and 68.25 %. When we run more iterations for DT, there is no clear trend of increase or decrease in UAR on the validation set. This suggests over-fitting. In this situation, accuracy on the validation set depends on randomness of the decision boundary in the hypothesis space, and the boundary margin is already too narrow.

On the other hand, the simpler DS model improves with more training steps. Its UAR score improves in both training set and validation set when iteration increases from 10 to 20. The accuracy on the training set is far below 100 %, but cannot be improved when iteration is over 20. DS reaches its upper bound of prediction power. We have seen that DS and DT both exhibit their best results on the Stroop Dual Task model, and there is no need to explore a more complex model structure. The fact

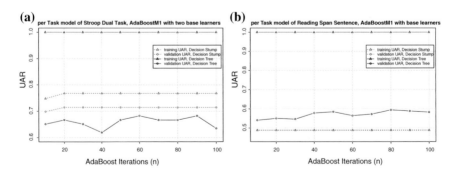

Fig. 1 Per task models of Stroop Dual Task (**a**) and Reading span task (**b**); AdaBoost.M1 with Decision Stump and Decision Tree base learners

that DS outperforms DT as an AdaBoost base learner is therefore to be expected. The sub-tasks with the smallest numbers of instances (Stroop dual, and Stroop time pressure) tend to favour simpler models that are less prone to overfitting.

However, DT outperforms DS as a base learner for AdaBoost.M1 in the Reading Span Sentence task (Table 3). DS training UAR remains below 50 % when training iterations increases from 10 to 100 (Fig. 1b). This is a sign of under-fitting, suggesting that DS cannot represent the variances in a Reading task with 825 instances (Table 1). As in the previous per task Stroop models, the DT based classifier's training UAR is 100 % when iteration equals 10, indicating that it does not suffer from the same problem. Unlike the previous case, however, in the reading task model, the UAR of DT on the validation set has a roughly increasing trend with more iterations. Prediction power is increasing with a more complex model, so here there is no indication of over-fitting. More iterations or more complex DT base learners could induce better UAR on the validation set.

4 Discussion

In this paper we proposed solutions for classifying three levels of objective load, with evidence of 6,374 speech features. In contrast to the rich feature set, there are only 1,838 instances spanning four different tasks. Since a moderately tuned SVM classifier only achieves a 44.4 % baseline on a Stroop task, our results serve to emphasise the importance of data cleansing and dimensionality reduction in this study.

We found that dimensionality reduction by feature extraction through PCA and DCT harms performance in boosting as well as other models. This may be due to the differences of mean values among the features and the lack of an effective unsupervised way of normalising these values on a per speaker basis. On the other hand, the supervised CfsSubsetEval filter proved to be an effective feature selection method. The features with high correlation with class variable and low inter-correlation with other features were favoured. Multicollinearity is thus alleviated in this large feature set. The reduced feature set mainly contains frequency signals (MFCC and F0) and sound quality measures (log HNR), instead of energy related features (RMS). The reduced feature set does improve accuracy and improves on the SVM baseline for the Stroop data (Table 2).

The outcome of feature selection is encouraging, but we still need to improve model accuracy by controlling the complexity of a supervised learning model. The boosting model combines the predictions from multiple classifiers and is generally more accurate than a single classifier. The training iterations act as a controller of model complexity. In the first round, a base classifier is built. In the next round, the weight of the $n + 1$ base learner is D_{n+1}, which is higher on instances that learner n has error on. The final decision is a collective vote by weighted N base learners. When boosting has no error on the training set, the generalisation power of the base learner is enough for the current input. When validation accuracy keeps

increasing with training accuracy stable at 100 %, it is necessary to try to model with
more iterations, thereby increasing the risk of over-fitting. However, when training
accuracy remains stable at low values as the number of iterations increases, there
is little point in proceeding. Such base learner is not complex enough to represent
feature variances adequately.

5 Conclusion

We presented an exploration of feature selection and modelling trade-offs to be taken
into account when approaching the challenge of categorising a speaker's cognitive
load state based on acoustic features. We found that while Frequency signals (MFCC
and F0) and sound quality measures (log HNR) are critical in determining the levels
of cognitive load, energy related features (RMS) seem contingent to this task.

Under appropriate settings of base learner complexity, the boosting classifier
exceeds a strong SVM baseline in most Stroop tests. However, the former proved less
effective in the reading span sentence tasks. This suggests that it may be necessary
to study cognitive load prediction differently for each setting.

This is, however, a complex challenge and as the results reported here demonstrate,
there is ample room for further exploration. In the near future we plan to investigate
unsupervised ways of normalising the features per speaker as well as explore models
that can take advantage of global data in per task modelling.

References

1. Ahmed, N., Natarajan, T., Rao, K.R.: Discrete cosine transform. IEEE Trans. Comput. 23(1),
   90–93 (1974)
2. Bouamrane, M.M., King, D., Luz, S., Masoodian, M.: A framework for collaborative writing
   with recording and post-meeting retrieval capabilities. IEEE Distrib. Syst. Online (2004)
3. Campbell, N.: On the use of nonverbal speech sounds in human communication. In: Esposito,
   A., et al. (eds.) Verbal and Nonverbal Communication Behaviours. Lecture Notes in Computer
   Science, pp. 117–128. Springer, Berlin (2007)
4. Conway, A.R., Kane, M.J., Bunting, M.F., Hambrick, D.Z., Wilhelm, O., Engle, R.W.: Working
   memory span tasks: a methodological review and users guide. Psychon. Bull. Rev. 12(5), 769–
   786 (2005)
5. Gevins, A., Smith, M.E.: Neurophysiological measures of cognitive workload during human-
   computer interaction. Theor. Issues Ergon. Sci. 4(1–2), 113–131 (2003)
6. Hall, M., Frank, E., Holmes, G., Pfahringer, B., Reutemann, P., Witten, I.H.: The WEKA data
   mining software: an update. ACM SIGKDD Explor. Newsl. 11(1), 10–18 (2009)
7. Hall, M.A.: Correlation-based Feature Subset Selection for Machine Learning. The University
   of Waikato, Hamilton (1999)
8. Luz, S.: The non-verbal structure of patient case discussions in multidisciplinary medical team
   meetings. ACM Trans. Inf. Syst. 30(3), 17:1–17:24 (2012)
9. Luz, S., Su, J.: The relevance of timing, pauses and overlaps in dialogues: detecting topic
   changes in scenario based meetings. In: Proceedings of Interspeech 2010. pp. 1369–1372.
   ISCA, Chiba, Japan (2010)

10. Roy, D.M., Luz, S.: Audio meeting history tool: interactive graphical user-support for virtual audio meetings. In: Proceedings of the ESCA workshop: accessing information in spoken audio. pp. 107–110. Cambridge University, April 1999
11. Schuller, B., Steidl, S., Batliner, A., Epps, J., Eyben, F., Ringeval, F., Marchi, E., Zhang, Y.: The Interspeech 2014 computational paralinguistics challenge: cognitive & physical load. In: Proceedings of Interspeech 2014. ISCA (2014)
12. Stroop, J.R.: Studies of interference in serial verbal reactions. J. Exp. Psychol. **18**(6), 643–662 (1935)
13. Unsworth, N., Heitz, R.P., Schrock, J.C., Engle, R.W.: An automated version of the operation span task. Behav. Res. Methods **37**(3), 498–505 (2005)
14. Yap, T.F.: Speech production under cognitive load: Effects and classification. Ph.D. thesis, The University of New South Wales (2012)

Unit Selection Using Acoustic Supra-Segmental Cues to Improve Prosody

Anjana Babu and Anil K. Sao

Abstract Improving prosody of synthesized speech is a very challenging task. In this paper, we propose an approach for improving the prosody by making use of acoustic supra-segmental cues for selecting units (syllables) in unit selection based speech synthesis (USS). It is based on the observation that certain acoustic features exhibit consistency at phrase level. This is an improvement of the method proposed in our earlier work (Babu et al. Twentieth National Conference on Communications (NCC), 2014 [1]), where units are selected based on the likelihood of the acoustic similarity of the adjacent units. The proposed approach is language independent and is evaluated using five Indian languages. The results show that the synthesized speech is quite natural.

1 Introduction

Incorporating prosody in unit selection based speech synthesis systems (USS) is a very challenging problem. The existing methods involve estimating a target prosody for the text and selecting units to match the target prosody. Traditionally, linguistic information is used to predict prosody. For example, part-of-speech (POS) is used for predicting the prosody in English [15]. The major challenge of such approaches is the lack of availability of sufficient linguistic resources for many languages. In USS systems also, prosody is first predicted using linguistic information from the text, followed by selecting units with the target prosodic features such as f_0 [8] or the spectral characteristics, f_0 and duration of each phone class [13], etc. There are also approaches where prosody is applied as a post processing step using HNMs and PSOLA based techniques [3]. The inherent problem with prosody

A. Babu (✉) · A.K. Sao
School of Computing and Electrical Engineering, Indian Institute of Technology Mandi, Mandi, India
e-mail: anjana_babu@alumni.iitmandi.ac.in

A.K. Sao
e-mail: anil@iitmandi.ac.in

© Springer International Publishing Switzerland 2016
A. Esposito et al. (eds.), *Recent Advances in Nonlinear Speech Processing*,
Smart Innovation, Systems and Technologies 48,
DOI 10.1007/978-3-319-28109-4_27

modification techniques is that they result in the loss of naturalness [4]. Hence, the use of such approaches nullify the advantages of using natural speech segments in USS based systems. For TTS for Indian languages, predicting prosody using linguistic information is not very practical as there are many languages with very little linguistic resources. There are some approaches for predicting certain acoustic features using machine learning techniques for incorporating prosody in Indian languages [9, 11, 12]. But most of them involve modifying the signal to the desired prosody.

In this paper, we are using acoustic features to improve the prosody of USS based systems without using linguistic information. This is an extension of our previous work, where units are selected in such a manner that the differences in the acoustic features of adjacent units are minimized to improve the naturalness of synthesized speech [1] (See Sect. 3). In the proposed approach, each phrase is considered as a prosodic entity and units are selected in such a manner that the difference in acoustic features of syllables, viz., average f_0 and average energy of syllables, are consistent within a phrase. Also, in order to maintain the speaking rate of the entire utterance, the difference in the duration of syllables within phrase and at phrase boundaries are not allowed to vary too much. The difference in acoustic features are modeled by Gaussian distributions for all the locations of the syllables in the utterance, which is explained in detail in [1].

The rest of the paper is organized as follows. Section 2 details the speech database used. Section 3 describes an existing approach for selecting units in unit selection framework. Section 4 describes the approach used in this paper for predicting phrase breaks. Section 5 explains the proposed approach. Section 6 gives the experimental results and Sect. 7 concludes the work.

2   Speech Database Used

The performance of the proposed approach is evaluated using the speech corpus given in Table 1. The data in the database consists of recordings of declarative sentences, which are referred to as utterances. The utterances are segmented at syllable level using a semi-automatic labeling tool called DonLabel [6]. The phrases are marked manually in these utterances.

Table 1 Language databases used

Language	Hours of data	Speaker	Language family
Hindi	6.5	Male	Aryan
Tamil	5.0	Female	Dravidian
Gujarati	6.8	Male	Aryan
Marathi	4.2	Male	Aryan
Rajasthani	3.3	Female	Aryan

3 Unit Selection Using Acoustic Features

In [1], a method to select units based on the acoustic similarity was discussed. The units for synthesis are selected in such a manner that the likelihood of the difference in acoustic features is maximized. The difference in acoustic features, viz. duration, average f_0 and average energy of syllables, are modeled using Gaussian distributions, most of which have mode around zero. This results in selecting units that have minimal variations with respect to these acoustic features. The likelihood of two units getting selected are obtained as scores based on the difference in the acoustic features of the adjacent candidate units.

The units are selected by first computing the scores for every pair of candidates, followed by a backtracking algorithm to select units based on the maximum scores obtained by the pairs of candidates. This approach focuses mainly on improving the naturalness across the entire utterance and does not necessarily improve the prosody. In the approach proposed in the paper, units are selected using this method.

4 Phrase Break Prediction

In the proposed approach, it is essential to predict phrase breaks in the given text. For predicting phrase breaks, case markers [2] and word terminal syllables [16] are used. The phrase breaks are predicted using CART (Classification and Regression tree), which uses features related to case markers (words) or word terminal syllables (syllables). In this work, word-terminal syllables are used in Dravidian languages considering the agglutinative nature. For Aryan languages, case markers or word-terminal syllables can be used, as both of them gave similar results.

5 Unit Selection Using Acoustic Features and Supra-Segmental Cues

Prosody is the rhythm and meter of speech [14]. It is considered as a supra-segmental characteristic of speech [10]. Some of the major acoustic features that characterizes prosody are f_0, energy and duration of speech [7]. It is important to incorporate prosody in synthesized speech because they render the speech more natural and meaningful [5]. Since prosody exhibits variability to some extent, no target values are computed for the acoustic features. Setting target values often result in selecting some units that are off the mark due to the unavailability of units, and hence, degrades the quality of the synthesized speech. The proposed approach selects units such that fairly decent prosody can be obtained from the units available in the database. Also, the proposed approach ensures the consistency in the difference of certain acoustic features at the phrase level only and not at the utterance level. This has an additional

advantage that, since the number of syllables in a phrase is less, the chances of
selecting syllables with acoustic similarities is higher. The main idea is to avoid
imposing any constraints that might adversely affect the unit selection, as well as
select a sequence of syllable units that result in good prosody. The proposed approach
makes use of the information about phrases for selecting units. The units are selected
using the scores mentioned in Sect. 3. However, instead of computing scores for
every pair of syllables in the utterance, phrase information is used to decide the
scores to be computed for each pair. Scores based on average f_0 and average energy
are omitted for adjacent syllables that belong to different phrases, whereas scores
based on duration is computed for every pair of syllables irrespective of their phrases.
Synthesizing speech using the proposed approach involves the following steps.

1. Predicting phrase breaks using decision trees, which are trained on contextual
   information about syllables.
2. Training statistical models describing the distribution of difference in acoustic
   features at various locations in the utterance.
3. Selecting units using phrase information such that the likelihood of the variation
   in acoustical features is maximized.

The proposed approach is depicted in Fig. 1. It can be seen that the first two steps
mentioned above are performed during the training stage and the last step during
the synthesis stage. In the synthesis stage, the phrase breaks are predicted using
decision trees. The scores for various adjacent candidate units are computed using
the Gaussian distributions of acoustic features for syllables at various locations. The
scores for average f_0, average energy and duration are computed if there is no phrase
break present between two syllables. In case there is a phrase break between two
syllables, the score for duration only is computed.

This approach is validated by the analysis of the acoustic features of the syllables in
natural utterances within phrases and utterances, which is discussed in the following
subsections.

Fig. 1 Block diagram
depicting the proposed
approach

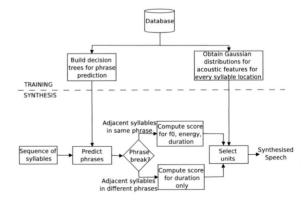

5.1 f_0 Contour

The deviation of average f_0 of syllables from the average value in phrase and in utterance were observed. The histograms corresponding to these observations are shown in Fig. 2a, b. It can be observed that the histogram for phrase exhibited unimodal distribution, with mode at 0. This suggests that the variation of f_0 within phrase is not large. However, the histogram for the difference in average f_0 from the average value of the corresponding utterance showed a bimodal distribution. This indicates that there is significant difference in the utterance. This is similar to the observation in [17] that, within a single breath group or phrase of simple sentences, the f_0 fluctuates between two abstract lines, the baseline and the plateau, and diminishes towards the end of the phrase. Also, there is a tendency to reset f_0 when a new phrase starts [17]. This can be observed in Fig. 3a, b.

The two sub figures are not marked (a) and (b), Please include it, the same way as in other figures.

5.2 Energy Contour

It was observed that all the phrases have roughly the same energy level across an utterance. Within a phrase, the energy of the last syllable tends to be slightly lower and the energy dies down rather slowly, compared to other syllables in the phrase. This can be observed in the last syllable of the last phrase in Fig. 3b. In many phrases, it can be observed that the energy is rather high at the beginning of the phrase (Fig. 3a). In most of the observations, the energy starts high and decreases towards the end of the phrase. The reason behind this could be that, after a pause, the speaker is able to

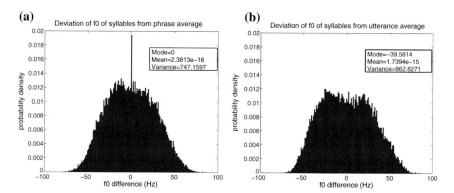

Fig. 2 Normalized histograms depicting **a** the difference in average f_0 of syllables from the average of average f_0 of all syllables in the corresponding phrase, **b** the difference in average f_0 of syllables from the average of average f_0 of all syllables in the corresponding utterance in Rajasthani database

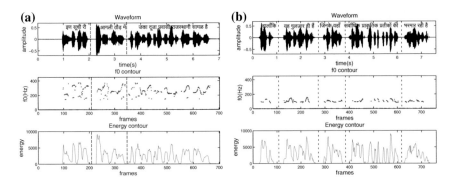

Fig. 3 The waveform, f_0 contour and energy contour of **a** a Rajasthani utterance containing 3 phrases, and **b** a Hindi utterance containing 5 phrases

start a new phrase with more energy which decreases towards the end of the phrase due to fatigue. The difference in energy between the last syllable of a phrase and the first syllable of the succeeding phrase is relatively high.

Certain deviations to the energy pattern discussed above can be observed. For example, in Fig. 3b, certain syllables within a phrase have relatively high energy with respect to the neighboring syllables. This could be because of emphasis, etc. in the speech. But, in general, it is observed that a consistent energy pattern is quite sufficient for a reasonably good prosody in the case of declarative speech.

5.3 Duration

The duration of the last syllables in phrases and utterances were observed to check for any tendency for final lengthening of syllables [17] and it was found that they did not deviate much from the average duration of syllables in the phrase or utterance in which they were present. In the database, there were no significant number of cases of variations in duration near prosodic boundaries and final lengthening of syllables. The duration of syllables did not vary much within an utterance and the variations were gradual.

6   Experimental Evaluation

The systems were built for five languages listed in Sect. 2. Different types of systems were built for each language:

(a) Type 1 with average f_0 across phrases and duration across utterance,
(b) Type 2 with average energy across phrases and duration across utterance, and

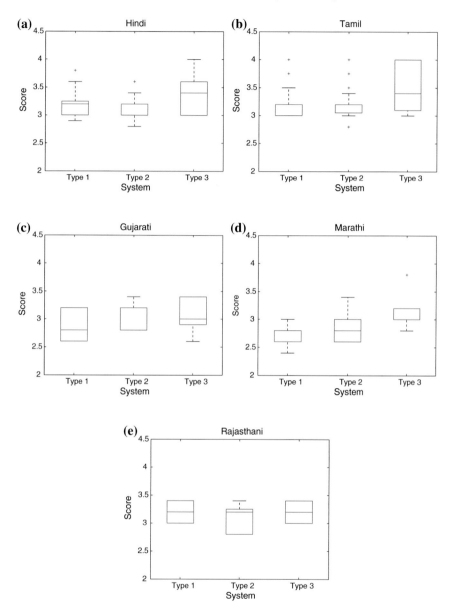

Fig. 4 Box plots for scores of various systems that use phrase information. **a** Box plot for Hindi systems. **b** Box plot for Tamil systems. **c** Box plot for Gujarati systems. **d** Box plot for Marathi systems. **e** Box plot for Rajasthani systems

(c) Type 3 with average f_0 and average energy across phrases and duration across utterance.

Utterances were synthesized from each of these three types of systems. These utterances were used for a subjective evaluation test called Degradation Mean Opinion Score (DMOS) test to assess the naturalness of the synthesized systems. The box plots corresponding to these scores are given below in Fig. 4. It can be observed that for most of the languages the median of system Type 3 is above the boxes of the other two systems. In the case of Marathi (Fig. 4d), the box for system Type 3 is above the boxes of the other two systems, indicating that Type 3 is better. This indicates that it is likely that the systems using f_0 and energy at phrase level is better. This could be because phrases are characterized by f_0 contour and energy pattern.

7 Conclusion

The proposed approach makes use of the variation in acoustic features with respect to the phrases in the utterance for incorporating prosody. It was observed that the average f_0 and average energy of syllables exhibited continuity within phrases, and not necessarily across utterances. But the duration of the syllables was consistent across the utterances. This observation was exploited for selecting units for incorporating prosody in unit selection framework. The results of the subjective evaluation of the speech synthesized using the proposed approach showed that both average f_0 and average energy are required, and not any one of them, for improving the prosody. The approach proposed here uses only acoustic information and not linguistic information for improving the prosody of speech. Since this approach is language independent and has already been tested on five Indian languages, this approach can be used for other languages.

Acknowledgments The authors express their sincere gratitude to Prof. Hema A Murthy and Raghava Krishnan K for the valuable suggestions and discussions. Also the authors would like to thank DeitY for funding the project, *Development of Text to Speech Systems for Indian Languages* (11(7)2011-HCC(TDIL)).

References

1. Babu, A., Krishnan, K.R., Sao, A., Murthy, H.: A probabilistic approach to selecting units for speech synthesis based on acoustic similarity. In: Twentieth National Conference on Communications (NCC), pp. 1–6 (2014)
2. Bellur, A., Narayan, K.B., Krishnan, K., Murthy, H.A.: Prosody modeling for syllable-based concatenative speech synthesis of Hindi and Tamil. In: National Conference on Communications (NCC), 2011, pp. 1–5. IEEE (2011)
3. Beutnagel, M., Conkie, A., Syrdal, A.K.: Diphone synthesis using unit selection. In: The Third ESCA/COCOSDA Workshop (ETRW) on Speech Synthesis (1998)

ational Conference on Speech Communication and Technology (INTER-SPEECH/Eurospeech). pp. 81–85. Lisbon, Portugal (2005)

4. Campbell, N., Black, A.W.: Prosody and the selection of source units for concatenative synthesis. In: Progress in speech synthesis, pp. 279–292. Springer (1997)
5. Cutler, A., Dahan, D., Van Donselaar, W.: Prosody in the comprehension of spoken language: a literature review. Lang. Speech **40**(2), 141–201 (1997)
6. Deivapalan, P., Jha, M., Guttikonda, R., Murthy, H.A.: Donlabel: an automatic labeling tool for Indian languages. Energy **2**, 4 (2008)
7. Dutoit, T.: An Introduction to Text-to-Speech Synthesis. Kluwer, Dordrecht (1997)
8. Fujisawa, K., Campbell, N.: Prosody-based unit selection for Japanese speech synthesis. In: The Third ESCA/COCOSDA Workshop (ETRW) on Speech Synthesis (1998)
9. Krishna, N.S., Murthy, H.A.: Duration modeling of Indian languages Hindi and Telugu. In: Fifth ISCA Workshop on Speech Synthesis (2004)
10. Nooteboom, S.: The prosody of speech: melody and rhythm. Handb. Phon. Sci. **5**, 640–673 (1997)
11. Rao, K.S.: Application of prosody models for developing speech systems in Indian languages. Int. J. Speech Technol. **14**(1), 19–33 (2011)
12. Reddy, V.R., Rao, K.S.: Two-stage intonation modeling using feedforward neural networks for syllable based Text-to-Speech synthesis. Comput. Speech Lang. **27**(5), 1105–1126 (2013)
13. Sakai, S., Shu, H.: A probabilistic approach to unit selection for corpus-based speech synthesis. In: 9th European Conference on Speech Communication and Technology (INTERSPEECH/Eurospeech). pp. 81–85. Lisbon, Portugal (2005)
14. Studdert-Kennedy, M.: Speech perception. Lang. Speech **23**(1), 45–66 (1980)
15. Taylor, P., Black, A.W.: Assigning phrase breaks from part-of-speech sequences. Comput. Speech Lang. **12**(2), 99–117 (1998)
16. Vadapalli, A., Bhaskararao, P., Prahallad, K.: Significance of word-terminal syllables for prediction of phrase breaks in Text-to-Speech systems for Indian languages. In: 8th ISCA Workshop on Speech Synthesis. pp. 209–214. Barcelona, Spain (August 2013)
17. Vaissière, J.: Language-independent prosodic features. Prosody: Models and Measurements, pp. 53–66. Springer, Berlin (1983)

A User-Centric Design of Service Robots Speech Interface for the Elderly

Ning Wang, Frank Broz, Alessandro Di Nuovo, Tony Belpaeme
and Angelo Cangelosi

Abstract The elderly population in the Europe have quickly increased and will keep growing in the coming years. In facing the elder care challenges posed by the amount of seniors staying alone in their own homes, great efforts have been made to develop advanced robotic systems that can operate in intelligent environments, and to enable the robot to ultimately work in real conditions and cooperate with elderly end-users favoring independent living. In this paper, we describe the design and implementation of a user-centric speech interface tailored for the elderly. The speech user interface incorporating the state of the art speech technologies, is fully integrated into application contexts and facilitates the actualization of the robotic services in different scenarios. Contextual information is taken into account in the speech recognition to reduce system complexity and to improve recognition success rate. Under the framework of the EU FP7 Robot-Era Project, the usability of the speech user interface on a multi-robots service platform has been evaluated by elderly users recruited in Italy and Sweden through questionnaire interview. The quantitative analysis results show that the majority of end-users strongly agree that the speech interaction experienced during the Robot-Era services is acceptable.

Keywords Human-robot interaction · Service robots · Elderly users · Speech interface

N. Wang (✉) · F. Broz · A. Di Nuovo · T. Belpaeme · A. Cangelosi
Centre for Robotics and Neural Systems, Plymouth University, Plymouth, UK
e-mail: ning.wang@plymouth.ac.uk

F. Broz
e-mail: frank.broz@plymouth.ac.uk

A. Di Nuovo
e-mail: alessandro.dinuovo@plymouth.ac.uk

T. Belpaeme
e-mail: tony.belpaeme@plymouth.ac.uk

A. Cangelosi
e-mail: a.cangelosi@plymouth.ac.uk

A. Di Nuovo
Faculty of Engineering and Architecture, University of Enna Kore, Enna, Italy

© Springer International Publishing Switzerland 2016
A. Esposito et al. (eds.), *Recent Advances in Nonlinear Speech Processing*,
Smart Innovation, Systems and Technologies 48,
DOI 10.1007/978-3-319-28109-4_28

275

1 Introduction

1.1 *Robots for the Elderly*

The increasing ageing population leads to a growing burden of elder care world-wide, which is especially true for the developed countries like those in the Europe. According to an European Union report, the number of people elder than 65 years old was 87 million in the Europe in 2010 [2]. It is estimated by the World Health Organization (WHO) that the elder population over the age of 60 is expected to be around 2 billion in 2050 [20]. One of the problems caused by the growing older population is that most of them still want to live in their own homes and to lead independent living as long as possible. With their gradually decaying physical and cognitive abilities, smart environments and assisting facilities, such as housekeeping, mobility support, social communication, and reminding systems are needed in their living places. In this circumstance, socially assistive robotic (SAR) platforms designed for improving independent living and caring for the elderly users, which can provide various services in both at-home and outdoor environments, are very much desired. For example, Robot *Robear*, which is developed by the Japanese robotics company RIKEN has human-like limbs to help move and carry objects. Others like the Aldebaran Robot *Pepper*, with emotion intelligence equipped, are designed to offer therapeutic care to the user. Meanwhile, elder adults living independently have also expressed their willingness to have robots live with them at home [17]. In the elder care domain, SAR platforms are usually integrated in an Ambient Assisted Living (AAL) environment [13], in which smart assistance systems and personal robots are designed and developed for a safer quality life at home.

1.2 *Robotic User Interface*

In order to achieve certain social, cognitive, and task outcomes goals in human-robot interaction (HRI), robots are needed to display appropriate social behaviours [6]. In realizing a SAR platform, one of the most challenging aspects regarding HRI is social communication. User interfaces in SAR can be keyboard, touch screen, gesture, and natural language, etc. Natural language based technologies such as automatic speech recognition (ASR), text-to-speech (TTS) synthesis, and language understanding have been evidently advanced in the past years, which can be seen from the great success gained by speech and language based technology products in consumer electronic markets, such as Siri[1] by Apple and Now[2] by Google. Speech technology is viewed as a major interaction modality in many application domains, for examples, customer

---

[1] http://www.apple.com/ios/siri/.

[2] http://www.google.com/landing/now/.

schedule information query and booking systems over telephone have employed voice-based interface for flights [8], trains [10] and restaurants and hotel booking [12]. At the same time, speech interface has also been widely engaged in multi-modal user interactive systems, such as those in smart homes [14] and AAL [7]. It has been found that among all human–machine interaction media, speech interaction is the one most accepted by users, especially for elderly people [14]. However, challenges exist in developing user-centric and high performance speech interaction tools. It is known that large vocabulary continuous speech recognition (LVCSR) is always time-consuming, substantial efforts such as data collection, model training, user adaptation, and parameter tuning are needed before actual deployment. LVCSR with state of the art recognition protocols and algorithms still report word error rates of around 20 % on average [15, 16], or of 10 % if trained for a more specific domain [19]. Accents, dialects and the mixed usage of multiple languages cause other failures in recognition. In domain-specific interaction tasks, contextual cues could be useful for enhancing speech recognition performance in either HRI or AAL cases. The central idea of context-sensitive speech recognition is to associate different contexts or dialogue states with individual language models or more specifically, grammars. In this case, grammar switching is indicated based on dialogue movement. This method has been shown led to more robust speech recognition [11]. Lemon showed in [11] the efforts on context-sensitive speech recognition in a dialogue system with more flexible and effective grammar switching strategy. Contextual information was also used to analyze the humans engagement towards the robot while using the dialogue system [9].

In this paper, we describe the speech interface employed in our elder-robot inter-action investigation. At first, the elder service robotic platform developed in the EU FP7 Robot-Era Project is introduced. Secondly, an overview of the speech-based user interface deployed is given, in which the context-aware grammar-switching mechanism for speech recognition efficacy is highlighted. After that, preliminary results of HRI experiments on a series of the Robot-Era tasks aiming at evaluating the speech-based interface are shown. Finally, conclusive remarks of this investigation are gained.

2 Elder Service Robotic Platform in Robot-Era

The EU FP7 Project Robot-Era[3] [1] aims at integration and implementation of advanced robotic systems including SAR and AAL architectures, with an ultimate goal to provide intelligent environments and facilities in real scenarios for the ageing population. To this end, the general feasibility, scientific/technical effectiveness and social/legal acceptability of the package of robotic services offered as well as the smart environments where the robot is operating in, is assessed by real users in actual

---

[3]http://www.robot-era.eu/.

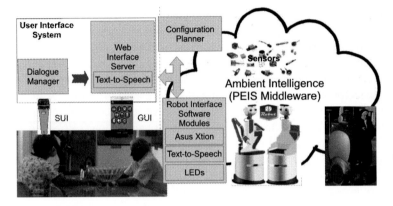

Fig. 1 Multi-robots platform and ambient intelligence architecture of the Robot-Era Project [1]

scenarios. The end users of this system are elderly people leading independent living in their daily life. In this project, several already available and commercial robotic components are adapted and integrated in both indoor (e.g., domestic, condominium) and outdoor environments to ensure independent, comfortable and safe living quality for the ageing population.

The Robot-Era services are specially designed to meet the needs of independently living elderly people in various scenarios ranging from indoor house keeping to outdoor walking support. Studies indicate that elderly people favor multi-modal user-robot interaction [18]. Among them, speech and gestures have been found most preferable by them [4]. Considering the fact that touch screen based mobile phone interface has gradually been accepted by more and more elderly users [3], in Robot-Era Project, two interfaces: graphic user interface (GUI) and speech user interface (SUI) are provided. Each of them can play as a sole-modal user interface, or they can work together as a dual-modal HRI platform. Figure 1 gives a whole picture of the multi-robots and ambient intelligent system architecture developed in the Robot-Era Project.

3 Speech Interaction

The SUI designed for the elderly people manipulating service robots is user-centric and domain-specific. We make efforts to provide a multilingual spoken dialogue system in various real-world scenarios. The spoken dialogue system employed is mainly composed of the following components: commercialized Nuance[4] speech

---

[4]http://www.nuance.com/.

recognizer and parser, open-source Olympus[5] dialogue manager, and Acapela[6] voice-as-a-service (VAAS) speech synthesizer. The Ravenclaw-based dialogue manager simplifies the authoring of complex dialogs, having general support to handle speech recognition errors, and is extendable to multi-modal input/output design [5]. The speech recognition and parsing is based on service-tailored grammars. To operate the SUI in real-world scenarios, one of the most important issues is to provide satisfactory speech recognition accuracy as well as to manage the complexity and variety of speech-based interaction. Considering service-specific spoken contexts might apply to the SUI, we propose to handle this issue from the following aspects: (1) developing context-aware language models (i.e., grammars in this study) to help reduce speech recognition complexity [11]; (2) achieving continuous and flexible dialogue flow by switching among different tasks without restarting the speech interaction module. Another significant issue in SUI is error handling. These errors usually result from speech recognition failures or misunderstandings due to false positives. The Olympus dialogue manager uses an error handling policy based on repeated prompts for recognition failures and explicit confirmation to ground recognized concepts in order to manage these sources of error. To employ context-aware speech recognition in a dialogue system, we load grammars dynamically according to the context change of verbal interaction. At the beginning of a speech interaction, the dialogue could only be initiated by the user via saying the wake-up word, which is defined to be the name of the robot in this work. A grammar containing the full list of all available services will be loaded immediately. After a specific service is chosen by the user through speech, which can either be a short command or a complete sentence, the dialogue manager will indicate the engine to switch to the according service-specific grammar. In manipulating with the contextual information, real-time SUI operation is ensured by avoiding complex large vocabulary language models, meanwhile, more robust speech recognition performance is achieved.

Our SUI is fully employable on the Robot-Era multi-robots platform. The speech interaction mode is available for all Robot-Era services, which include *communication* (via mobile phone or skype), *shopping, cleaning, food delivery, indoor escort, object manipulation, garbage collection, laundry, reminding, surveillance,* and *mobility support*. These indoor or outdoor services are fully supported in three languages: English, Italian, and Swedish. Figure 2 shows the dialogue flow path in a real SUI operation. In this example, the Robot-Era services *food delivery, shopping, laundry,* and *garbage collection* are asked by the user sequentially. In each dialogue stage, a specific grammar is enabled. Transitions from one grammar to another are indicated by the dialogue manager. It is noted that before each dialogue movement, a confirmation with the user is made to avoid wrong action caused by speech recognition failures.

---

[5]http://wiki.speech.cs.cmu.edu/olympus/index.php/Olympus/.
[6]http://www.acapela-group.com/.

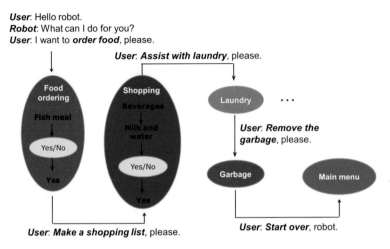

User: Hello robot.
Robot: What can I do for you?
User: I want to ***order food***, please.

User. ***Assist with laundry***, please.

Food ordering

Fish meal

Yes/No

Yes

Shopping

Beverages

Milk and water

Yes/No

Yes

Laundry   . . .

User. ***Remove the garbage***, please.

Garbage

Main menu

User. ***Make a shopping list***, please.

User. ***Start over***, robot.

Fig. 2 Robot-Era services manipulated by speech

4   Preliminary Experiments on Usability

To evaluate the usability of our SUI by elderly users, we designed and conducted a series of tests on selected Robot-Era services. For the services *shopping*, *reminding*, *indoor escort*, and *garbage collection*, the experiments were carried out on our Italian test site in Peccioli, Italy. The services *laundry* and *food delivery* were tested on the Swedish test site in Angen, Sweden. For the last service *communication*, tests were conducted on both sites. At this stage, we recruited 35 Italian subjects (22 females and 13 males), of an average age of 73.80 ± 5.81, and 12 Swedish subjects (4 females and 8 males) aged averagely at 70.67 ± 5.37 to participate in the HRI experiments.

During the experiments, subjects weren't instructed step-by-step on using the SUI, instead, they were shown a demonstration for each task. For further investigation, their spontaneous speech interaction with the robot were recorded. A questionnaire was completed by each subject. The questionnaire consisted of several items, and two of them were about the usability of the SUI. Subjects were asked to use five-point Likert type scale (1. strongly disagree; 2. disagree; 3. no opinion; 4. agree; 5. strongly agree) to answer the questions. Figure 3 shows the median scores valued by all the subjects for the two questions, each in a radar plot. It is observed that most of the subjects were satisfied with the SUI performance in the five services: *shopping*, *indoor escort*, *food delivery*, *garbage collection*, and *reminding*. For the other two services: *communication* and *laundry*, they still accepted it. In Fig. 4, the service-wise acceptability scores are described by an error bar plot. It is found that among the subjects, the SUI in all services get average scores above four out of five. In specific, the elderly users were more satisfied with the SUI experience for the four services: *shopping*, *reminding*, *garbage collection*, and *indoor escort* during the experiments.

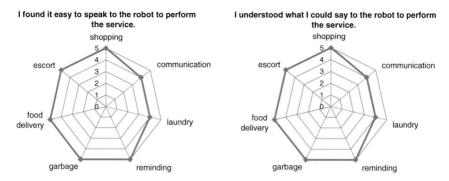

Fig. 3 User attitude towards the SUI under different application scenarios

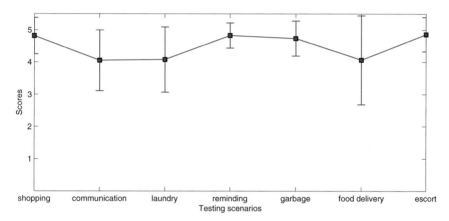

Fig. 4 Error bar plot of user acceptability scores in different testing scenarios by questionnaire interviews

5 Conclusions

The increasing ageing population in need of personal care and services causes huge pressure to the European countries, which turns out motivate the development of personal robots nowadays. One of the most challenging aspects for personal robots is social communication, which is especially difficult in the case of elder-robot interaction. In this paper, we introduce the design and implementation of a speech user interface specially tailored for elderly users. Although being developed under the framework of the EU FP7 Robot-Era Project, this interface can easily apply to other robotic platforms or used alone. The speech user interface is realized by a spoken dialogue system composed of the speech recognizer, parser, dialogue manager, and speech synthesizer, etc. To achieve low-complexity yet user-centric speech interaction system, we employ a context-sensitive speech recognition approach to enable flexible dialogue movement in real time. Evaluation made by elderly users in Italy and Sweden shows that for the tested Robot-Era services, subjects totally agree that the speech interaction experience during operating the robot is acceptable. Further elder-robot interaction investigation is ongoing, it is therefore expected that more findings on elder-centric speech interface will be gained in future.

Acknowledgments This work is fully supported by the European Union Seventh Framework Programme (FP7/2007-2013) grant no. 288899.

References

1. Robot-Era project: Implementation and integration of advanced robotic systems and intelligent environments in real scenarios for the ageing population. In: FP7-ICT-Challenge 5: ICT for Health, Ageing Well, Inclusion and Governance. Grant agreement number 288899
2. Active ageing and solidarity between generations. European Union (2011)
3. Al-Razgan, M., Al-Khalifa, H., Al-Shahrani, M., Al-Ajmi, H.: Touch-based mobile phone interface guidelines and design recommendations for elderly people: a survey of the literature. In: Lecture Notes in Computer Science, vol. 7666, pp. 568–574. Springer, Berlin Heidelberg (2012)
4. Arch, A., Abou-Zahra, S., Henry, S.: Older users online: WAI guidelines address the web experiences of older users. User Experience Mag. **8**(1), 18–19 (2009)
5. Bohus, D., Rudnicky, A.: The ravenclaw dialog management framework: architecture and systems. Comput. Speech Lang. **23**(3), 332–361 (2009)
6. Breazeal, C., Kidd, C., Thomaz, A., Hoffman, G., Berlin, M.: Effects of nonverbal communication on efficiency and robustness in human-robot teamwork. In: Proceedings of IEEE/RSJ International Conference on Intelligent Robots and Systems, pp. 708–713 (2005)
7. Di Nuovo, A., Broz, F., Belpaeme, T., Cangelosi, A., Cavallo, F., Esposito, R., Dario, P.: A web based multi-modal interface for elderly users of the Robot-Era multi-robot services. In: Proceedings of IEEE International Conference on Systems, Man and Cybernetics, pp. 2186–2191 (2014)
8. Karpov, A., Ronzhin, A., Leontyeva, A.: A semi-automatic wizard of oz technique for Let'sFly spoken dialogue system. In: Lecture Notes in Computer Science: Text, Speech and Dialogue, pp. 585–592. Springer (2008)
9. Kipp, A., Kummert, F.: Dynamic dialog system for human robot collaboration—Playing a game of pairs. In: Proceedings of 2nd International Conference on Human-Agent Interaction, pp. 225–228 (2014)
10. Lamel, L., Rosset, S., Gauvain, J., Bennacef, S.: The LIMSI arise system for train travel information. In: Proceedings of IEEE International Conference on Acoustics, Speech and Signal Processing, pp. 501–504 (1999)
11. Lemon, O.: Context-sensitive speech recognition in information-state update dialogue systems: results for the grammar switching approach. In: Proceedings of Eighth Workshop on the Semantics and Pragmatics of Dialogue, pp. 49–55 (2004)
12. Lemon, O., Georgila, K., Henderson, J., Stuttle, M.: An ISU dialogue system exhibiting reinforcement learning of dialogue policies: generic slot-filling in the TALK in-car system. In: Proceedings of Eleventh Conference of the European Chapter of the Association for Computational Linguistics, pp. 119–122 (2006)
13. Mayer, P., Beck, C., Panek, P.: Examples of multimodal user interfaces for socially assistive robots in ambient assisted environments. In: Proceedings of IEEE International Conference on Cognitive Infocommunications, pp. 401–406 (2012)
14. Portet, F., Vacher, M., Golanski, C., Roux, C., Meillon, B.: Design and evaluation of a smart home voice interface for the elderly: acceptability and objection aspects. Pers. Ubiquit. Comput. **17**(1), 127–144 (2013)

15. Sainath, T., Mohamed, A., Kingsbury, B., Ramabhadran, B.: Deep convolutional neural networks for LVCSR. In: Proceedings of IEEE International Conference on Acoustics, Speech and Signal Processing, pp. 8614–8618 (2013)
16. Sainath, T., Peddinti, V., Kingsbury, B., Fousek, P., Ramabhadran, B., Nahamoo, D.: Deep scattering spectra with deep neural networks for LVCSR tasks. In: Proceedings of INTERSPEECH, pp. 900–904 (2014)
17. Smarr, C., Prakash, A., Beer, J., Mitzner, T., Kemp, C., Rogers, W.: Older adults preferences for and acceptance of robot assistance for everyday living tasks. In: Proceedings of Human Factors and Ergonomics Society Annual Meeting, pp. 153–157 (2012)
18. Tang, D., Yusuf, B., Botzheim, J., Kubota, N., Chan, C.: A novel multimodal communication framework using robot partner for aging population. Expert Syst. Appl. **42**(9), 4540–4555 (2015)
19. Weninger, F., Schuller, B., Eyben, F., Wöllmer, M., Rigoll, G.: A broadcast news corpus for evaluation and tuning of German LVCSR systems (2014). eprint arXiv:1412.4616
20. World Health Organization: 10 facts on ageing and the life course. http://www.who.int/features/factfiles/ageing/en/ (2014)

New Method for Finding Optimum Number of Characteristics to Classify Speakers by Age

Cristina Muñoz-Mulas, Rafael Martínez-Olalla, Pedro Gómez-Vilda,
Agustín Álvarez-Marquina and Luis Miguel Mazaira-Fernández

Abstract It is known that the amount of characteristics may be the bottleneck of a digital processing system. Finding a good method to detect which characteristics are the most important to identify a speaker would get better results with less characteristics. The classification of an adult speaker by their age is a big challenge since the adulthood is a long period without significant changes in voice. This study proposes a new method based on F-ratio, dispersion metric and also correlation between parameters to find a rank of features. A bootstrapping procedure determines the optimum number of characteristics within a feature vector to characterize a speaker. The results are compared with other non linear ranking methods. The proposed algorithm achieves a better performance in most cases.

Keywords Biometry · Boostrapping · Rankfeatures · Features selection

1 Introduction

Voice experiences several changes throughout the cycle of life. These changes are due to growing factors which are directly influenced by nervous and hormonal changes in the individual. As a result there are physical and physiological differences which affect the organs of the phonatory system. These changes are reflected in voice. According to the age period of a person and the changes occurred in his/her body, the speaker may be classified into a stage of voice related to age: childhood, puberty, adulthood or senescence. The ages that fall into each of these stages are approximate,

C. Muñoz-Mulas (✉) · R. Martínez-Olalla · P. Gómez-Vilda · A. Álvarez-Marquina ·
L.M. Mazaira-Fernández
Escuela Superior de Ingenieros Informáticos, Universidad Politécnica de Madrid,
Campus de Montegancedo s/n, 28660 Boadilla del Monte, Madrid, Spain
e-mail: ce.munoz@upm.es

R. Martínez-Olalla
e-mail: rnolalla@junipera.datsi.fi.upm.es

© Springer International Publishing Switzerland 2016   285
A. Esposito et al. (eds.), *Recent Advances in Nonlinear Speech Processing*,
Smart Innovation, Systems and Technologies 48,
DOI 10.1007/978-3-319-28109-4_29

since not all individuals undergo changes at the same time and they don't live in the same environment [1, 2].

Feature selection has become one of the "hot spots" for the studies that use data set with a large number of variables. In recent decades several articles have been published in reference to feature selection [3, 4]. However, due to the improvement of the technology and the access to the information, the number of variables used have increased, which implies new methods to find relevant features [5].

2 Materials and Methods

This study uses Albayzin database [6], which is divided into age groups as follows: Females (young aged from 18 to 30 (77 speakers)/medium, aged from 29 to 41 (39 sp)/senior, aged from 43 to 55 (36 sp)) and Males (young, aged from 18 to 30 (77 sp)/medium, aged from 30 to 40 (39 sp)/senior, aged from 41 to 55 (36 sp)).

As described in previous studies [7] the speech signal (voice) is decomposed into glottal pulse and vocal tract and each of these three signals (voice, vocal tract and glottal pulse) are processed to obtain a MFC vector of 52 characteristics. In order to simplify the process of feature selection, LTA vectors are composed gathering all the information of each speaker as is detailed in [7]. As a result, a speaker is represented by one unique vector.

The analysis of the characteristics concludes that a better classification would be done if male and females are separated in different groups. In Fig. 1 we can see the location of each speaker in the database when the first two canonical components (c1, c2) from MANOVA are represented.

As we can see in Fig. 1, the young and senior groups are the one which seem to be more separable in both cases, males and females.

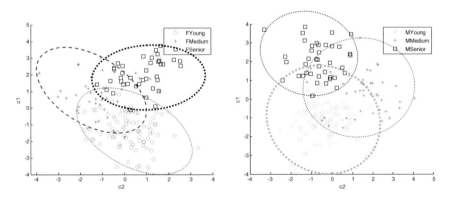

Fig. 1 Plot of males (*left*) and females (*right*) speakers represented by the first two canonical components obtained by MANOVA

We have assured that ANOVA conditions are guaranteed for most of the characteristics (or slightly violated). ANOVA + post-hoc tests are conducted to highlight the characteristics that are statistically relevant to separate age groups in both gender sets. Three commonly used post hoc test are compared (LSD: Least Significant Difference, HSD: Honest Significant Difference, G-H: Games-Howell, which doesn't assume homecedasticity). Results of this comparison are summarized in Figs. 2 and 3 for the 52 component vectors (20 voice, 12 glottal pulse, 20 vocal tract) where characteristics statistically significant for separating age groups are depicted

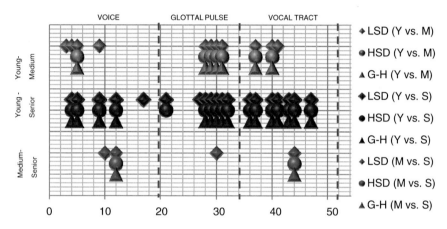

Fig. 2 Statistically significant characteristics for female speakers (Young vs. Medium, Young vs. Senior, Medium vs. Senior) for LSD, HSD and G-H tests

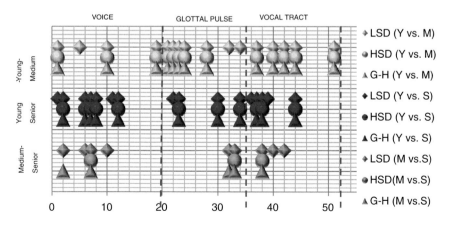

Fig. 3 Statistically significant characteristics for male speakers (Young vs. Medium, Young vs. Senior, Medium vs. Senior) for LSD, HSD and G-H tests

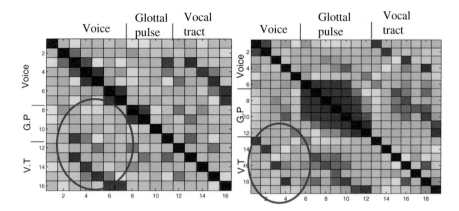

Fig. 4 Correlation matrix of males (*left*) and females (*right*). Dark red implies high correlation and dark blue high negative correlation

An interesting conclusion that can be inferred from Figs. 2 and 3 is that statistically significant characteristics that separate age groups in males are completely different from those that are able to separate age groups in females (see the alternation of relevant characteristics between both figures).

As LSD is the most relaxed method, we decide to include all the characteristics found by this test in our first feature vector. By doing so, we can reduce the initial 52 component vectors into 16 features vector for males and 19 feature vector for females.

Glottal pulse and vocal tract signals are highly decorrelated by the separation procedure and also are their respective MFCC vectors. However, they still have some correlation with voice characteristics. This remaining correlation is evident in Fig. 4, which shows the correlation matrix of the LSD selected characteristics for males and females. As we can see, only voice and vocal tract characteristics keep high correlation for some of their values.

Thus, selecting the characteristics using only F-ratio is not enough to avoid redundant information. This fact leads to the idea of creating a new ranking algorithm which uses not only F-ratio but also correlation and dispersion between groups. The criterion in the algorithm is described in Eq. 2 and is applied to the whole 52 characteristics initial vector introducing characteristics one by one and selecting in each step the variable which maximizes Cr.

$$Cr = \frac{\Sigma_{i=1}^{n} F_{c_i}}{D^{\alpha} \cdot (\Sigma \Sigma |corr_{gr}|)^{\beta}} \qquad (2)$$

where F_{c_i} is the F-ratio value of the ith included characteristic, n is the number of characteristics included to that point, D is the dispersion metric and $Corr_{gr}$ is the correlation matrix between characteristics included so far, α and β give different weights to the variables.

The procedure gives reordered 52-characteristic vector.

Dispersion

Dispersion (D), calculated by the Eq. (3) provides a measure to compare different sets of features. This measure represents the separation (distance) between the individuals within a group (compactness) in relation to the separation between groups. Thus, the smaller values of D, the greater interest.

As the number elements within each group may be different, this metric should be normalized as can be shown in the denominator of the expression in Eq. 3.

$$D = \frac{\frac{\frac{1}{n_1}\sum_{i=1}^{n_1}dist(p_i,c_1) + \frac{1}{n_2}\sum_{i=1}^{n_2}dist(q_i,c_2)}{dist(c_1,c_2)}}{\frac{\sqrt{n_1}\sqrt{n_2}}{\sqrt{n_1 + n_2}}} \tag{3}$$

where p and q represent a vector of the group 1 and 2 respectively, n_1 is the number of objects in group 1, n_2 the number of objects in group 2, c_1 is the centroid of the group 1 and c_2 is the centroid of the group 2.

Using different values for α and β in Cr (Eq. 2), several ranking methods and bootstrapping [8], we can generate several training/test sets, repeat the classification process a large number of times and extract statistical information about the distribution. In this case, we use LDA-diagonal linear function and SVM-RBF kernel methods to classify and extract statistical information about the use of different combinations to identify young/senior age.

We compare the results obtained by the refinement algorithm using different combinations of α and β and several ranking methods [5] as the ones provided by MATLAB "rankfeatures" function.

Description of the ordering algorithms compared
Refinement algorithm:

- Refcr1 → $\alpha = 1$ and $\beta = 1$
- Refcr2 → $\alpha = 1$ and $\beta = 1/2$
- Refcr3 → $\alpha = 2$ and $\beta = 1/2$
- Refcr4 → $\alpha = 1/2$ and $\beta = 1$

Ranking methods:

- t-test: Absolute value two-sample t-test with pooled variance estimate.
- entropy: Relative entropy, also known as Kullback-Leibler distance or divergence.
- Bhattacharyya: Minimum attainable classification error or Chernoff bound.
- ROC: Area between the empirical receiver operating characteristic (ROC) curve and the random classifier slope.
- Wilcoxon: Absolute value of the u-statistic of a two-sample unpaired Wilcoxon test, also known as Mann-Whitney.

The ranking algorithms can be tuned with a parameter alpha ranged between 0 and 1. Different values for that parameter have been tested, although the best results are obtained for high values of alpha (>0.9).

Bootstrapping methodology is applied to compare the results when LDA + diag linear function and SVM + RBF kernel function are used as classification methods. The number of test sets created is 20,000 in each case.

3   Results

After bootstrapping and classification with LDA and SVM, We obtain the results shown in Tables 1 and 2. Also a comparison is represented in Figs. 5 and 6, where we can also detect how many characteristics are good enough to represent speakers for age classification. As we can see, our method outperforms the rankfeatures functions, and this fact is more evident in the male case.

Increasing the feature vector item by item, n = 1 … 16 for males, n = 1 …19 for females, we can repeat this procedure and obtain the optimum number of characteristics for each ranking method In Figs. 5 and 6 the average of SVM and LDA classification error rates is shown (for males and females respectively). For males, we used ordered vectors from 3 to 16 characteristics whereas in females the vectors are from 3 to 19 characteristics (same number as LSD vector). These vectors are

Table 1 Average error rates after 20.000 iterations for different ranking algorithms with 16 characteristics. Males

	Ref_{cr1}	Ref_{cr2}	Ref_{cr3}	Ref_{cr4}	ttets	ent	bhatt	Roc	wilco
Mean									
LDA	0.277	**0.272**	0.276	0.274	0.289	0.290	0.292	0.290	0.305
SVM	**0.281**	0.284	0.286	0.288	0.300	0.297	0.302	0.290	0.330
AVG	**0.279**	**0.278**	0.281	0.281	0.295	0.293	0.297	0.290	0.317
Standard deviation									
LDA	0.063	0.063	0.062	0.063	0.062	0.063	0.065	0.065	0.069
SVM	0.063	0.062	0.062	0.062	0.061	0.061	0.063	0.063	0.066

Table 2 Average error rates after 20.000 iterations for different ranking algorithms with 19 characteristics. Females

	Ref_{cr1}	Ref_{cr2}	Ref_{cr3}	Ref_{cr4}	ttets	ent	bhatt	roc	wilco
Mean									
LDA	0.227	0.225	0.265	0.230	0.225	0.260	0.232	**0.221**	0.284
SVM	0.238	0.235	0.240	0.242	0.237	**0.221**	0.244	0.238	0.291
AVG	0.232	**0.230**	0.252	0.236	0.231	0.241	0.238	**0.230**	0.288
Standard deviation									
LDA	0.060	0.060	0.060	0.060	0.060	0.063	0.061	0.060	0.067
SVM	0.059	0.061	0.059	0.061	0.060	0.066	0.061	0.060	0.065

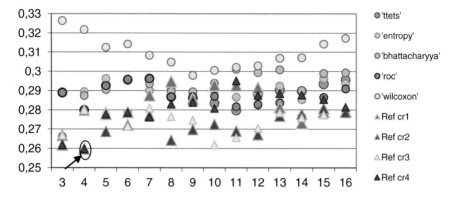

Fig. 5 Average of SVM and LDA classification error rates for males after bootstrapping as a function of the number of characteristics

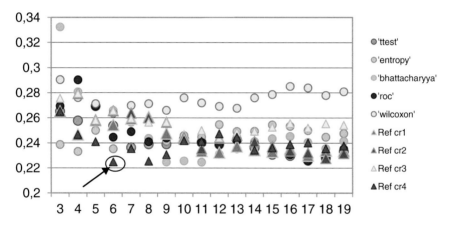

Fig. 6 Average of SVM and LDA classification error rates for female after bootstrapping as a function of the number of characteristics

Table 3 Average error rates for the optimum number of characteristics

	ttets	ent	bhatt	roc	Ref_{cr1}	Ref_{cr2}	Ref_{cr3}	Ref_{cr4}	Ref_{cr1}
Males (4)	0.2799	0.2894	0.2874	0.2799	0.3217	**0.2598**	0.2799	0.2799	**0.2598**
Females (6)	0.254	0.235	0.266	0.244	0.264	0.254	0.266	0.266	**0.225**

ordered by the nine ordering methods described above. As we can see, Refcr1 and Refcr4 achieve the best results in males using only 4 characteristics and Refcr4 gets the best result in females when 6 characteristics are used Thus, we would be able to reduce 16 and 19 feature vectors into 4 and 6 feature vectors.

To compare values, Table 3 summarizes the average error rates for females and males when the optimum number of characteristics is used in each ranking feature method.

4   Conclusion and Discussion

The study summarizes a deep process of analysis and search of feature selection in a challenging issue as age recognition by voice. Finally, the algorithm presented and the method based on bootstrapping enables us to identify characteristics to represent a speaker by en extremely low number of characteristics in a fast and easy way.

We have seen that female and male voices should be treated as different groups and also that the features which are important for one group are not for the other, so the groups must be treated separately. Some important conclusions may be detected from the analysis, and more study is needed to understand what is behind these differences.

The method proposed is a novel method based on F-ratio, correlation and dispersion metric which is able to sort a feature vector by the relevance of its characteristics avoiding redundancies to some extent. The comparison of the results of this method and other ranking methods reveals that our method outperforms all of them.

Also we can conclude that bootstrapping is a good way for finding the optimum number of characteristics as it can avoid the results to be conditioned by the selected training/test sets.

Acknowledgments   Work funded by grants TEC2012-38630-C04-04 and TEC2012-38630-C04-01, Plan Na-cional de I+D, Ministerio de Economía y Competitividad, Spain.

References

1. Harnsberger, J.D., Brown, W.S. Jr, Rothman, H., Hollien, H.: Speaking rate and fundamental frequency as speech cues to perceived age. J. Voice, The Voice Foundation **22**(1), 58–69 (2008)
2. Schötz, S.: Acoustic analysis of adult speaker age. In: Speaker Classification I, Lecture Notes in Computer Science, pp. 88–107. Springer (2007)
3. Blum, A., Langley, P.: Selection of relevant features and examples in machine learning. Artif. Intell. **97**(1–2), 245–271 (1997)
4. Kohavi, R., John, G.: Wrappers for feature selection. Artif. Intell. **97**(1–2), 273–324 (1997)
5. Guyon, I.: An introduction to variable and feature selection. J. Mach. Learn. Res. **3**, 1157–1182 (2003)
6. Casacuberta, F., García, R., Llisterri, J., Nadeu, C., Pardo, J.M., Rubio, A.: Desarrollo de un Corpus para la Investigación en Tecnologías del Habla. Procesamiento de Lenguaje Natural, Boletín no 12, 35–42 (1992)
7. Muñoz Mulas, C., Martínez Olalla, R., Gómez Vilda, P., Álvarez Marquina, A., Mazaira Fernánrez, L.M.: Discriminación de género basada en nuevos parámetros MFCC. In: I Workshop de tecnologías Multibiométricas para la identificación de Personas, pp. 22–16 (2010)
8. Efron, B.: Bootstrap methods: Another look at jackknife. Ann. Stat. **7**, 1–26 (1979)

Author Index

© Springer International Publishing Switzerland 2016
A. Esposito et al. (eds.), *Recent Advances in Nonlinear Speech Processing*,
Smart Innovation, Systems and Technologies 48,
DOI 10.1007/978-3-319-28109-4

Printed in the United States
By Bookmasters